Mathematical Reasoning™
Level B

Developing Math & Thinking Skills

Series Titles
Beginning Mathematical Reasoning™
Mathematical Reasoning™ Level A
Mathematical Reasoning™ Level B

Written By
Linda Brumbaugh
Doug Brumbaugh

Graphic Design By
Karla Garrett
Linda Laverty
Anna D. Chaffin
Doug Brumbaugh

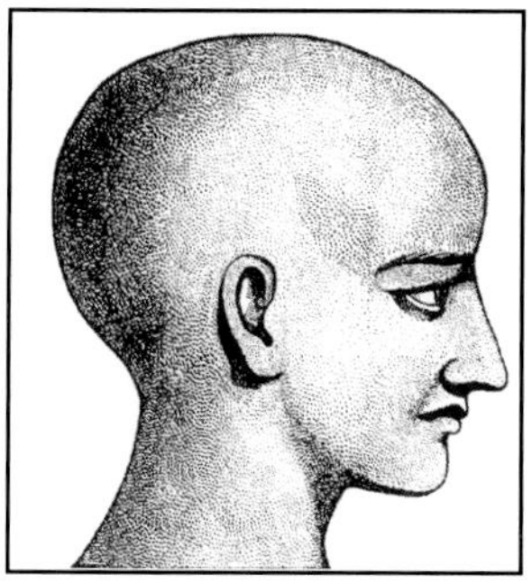

THE CRITICAL THINKING CO.™
(BRIGHT MINDS™)
www.CriticalThinking.com
P.O. Box 1610 • Seaside • CA 93955-1610
Phone 800-458-4849 • FAX 831-393-3277
ISBN 978-0-89455-885-6

Printed in China

ABOUT THE AUTHORS

Linda S. Brumbaugh

I retired after teaching a total of 31 years in grades three, four, and five. Both my BS from the University of Florida and Masters from the University of Central Florida are in Elementary Education. As I look back over my teaching career, I enjoyed seeing the excitement on the children's faces as they encountered new concepts, worked with a manipulative, experienced some new mathematical application, or played a new mathematical game. It was stimulating when they solved an intricate problem, discovered something new to them, or got caught up in some new mathematical trick. As they got excited about learning, so did I. Each day of every year brought some new learning opportunity for me and for the children. I continue to work with pre-school and elementary age children in the Sunday school system of our church. Our intent is to convey some of that excitement to each child who uses this book.

Douglas K. Brumbaugh

Depending on how you count, I have been teaching over 50 years. I taught in college, in-service, or K-12 almost daily. I received my BS from Adrian College and Masters and Doctorate in Mathematics Education from the University of Georgia. Students change, classroom environments change, the curriculum changes, and I change. The thoughts and examples used here are based on my teaching experiences over the years. The pages in this book are designed to spark the interest of each child who works with them.

TABLE OF CONTENTS

About the Authors ii

How to Use This Book v

NCTM STANDARDS / SKILLS	Number and Operations	Algebra	Geometry	Measurement	Data Analysis and Probability
Action		32, 144, 148	239, 259, 261	187	
Addition	4, 5, 8, 11, 12, 15, 19, 25, 27, 28, 30, 31, 34, 35, 37, 40, 46, 49, 54, 65, 74, 105, 107, 120, 134, 140, 142, 143, 145, 146, 147, 153, 155, 159, 161, 163, 166, 188, 190, 200, 202, 207, 208, 215 218, 252	52, 56, 59, 165	23	6, 7, 26, 57, 66, 67, 80, 82, 84, 262	
Bar Graph					73, 77, 117, 125, 150, 151, 157, 231, 236
Calendar				185, 240, 241, 258	117, 212, 231
Capacity	122, 126, 141			122, 124, 141, 179	231, 233
Count	1, 2, 4, 5, 8, 9, 10, 11, 12, 13, 19, 26, 27, 29, 30, 36, 49, 54, 60, 61, 65, 67, 70, 81, 83, 85, 93, 97, 101, 105, 107, 112, 128, 136, 137, 145, 146, 147, 153, 155, 164, 171, 174, 186, 188, 189, 202, 214, 220, 255	32, 52, 56, 59, 89, 102, 106, 114, 116, 118, 148, 165	38, 39, 42, 55, 211, 237	14, 18, 24, 26, 57, 66, 67, 71, 76, 80, 82, 84, 108, 115, 170, 232	73, 77, 127, 132, 133, 150
Fractions	17, 62, 63, 64, 178, 217		23, 43, 217		
Language	1, 3, 10, 21, 25, 28, 29, 31, 36, 54, 60, 62, 75, 97, 101, 112, 122, 136, 137, 158, 163, 173, 178, 246	32, 52, 59, 89, 114, 116, 165, 182, 251	16, 17, 23, 33, 39, 42, 44, 45, 48, 50, 55, 58, 149, 156, 162, 176, 211, 227	14, 51, 57, 103, 104, 108, 109, 110, 111, 113, 115, 122, 141, 152, 170, 179, 185, 199, 226, 229, 240, 241, 247, 262, 264	73, 77, 117, 119, 123, 125, 127, 132, 150, 151, 212, 233, 236
Length				103, 104, 152, 187, 203, 205, 210, 229	73, 77
Likelihood			50		127, 132, 133
Manipulative	27, 30, 54, 60, 61, 62, 65, 68, 81, 93, 145, 153, 155, 168, 172, 186, 188, 197, 202, 206, 214, 220	59, 102, 106, 118	96	66, 187	
Match	1, 3, 12, 19, 21, 29, 41, 47, 75, 92, 97, 121, 126, 131, 135, 139, 140, 146, 147, 167, 181, 191, 195, 200, 209, 214, 219, 223, 235, 245, 249, 257	89, 118	33, 48, 50, 55, 58, 156, 238, 250, 254	24, 25, 51, 80, 82, 84, 109, 110, 111, 126, 129, 141, 182, 204, 205, 206, 229, 263	

TABLE OF CONTENTS (Cont.)

NCTM STANDARDS (cont.)	Number and Operations	Algebra	Geometry	Measurement	Data Analysis and Probability
SKILLS (cont.) Money	163, 255	89, 182		6, 7, 14, 18, 24, 26, 57, 66, 67, 71, 78, 80, 82, 84, 129, 183, 197, 206, 232, 262	212
Number Line	34, 35, 37, 65, 83, 85, 159, 161, 175, 177, 246	102, 106, 118		198, 203, 205, 210	
Odd/Even	10, 97				
Order	2, 9, 29, 65, 97, 158, 164, 213, 228	32, 69, 72, 76, 79			
Pattern	2, 9, 13, 36, 64, 65, 70, 105, 107, 128, 158, 164, 189, 213, 228	20, 22, 32, 69, 72, 76, 79, 96, 100, 144, 154, 160, 192	33, 39, 42, 43, 48, 50, 55, 58, 162, 176, 239, 250, 254, 259, 261	24, 51, 82, 84, 141, 204, 230, 258	
Probability				113, 115	127, 132, 133, 224
Real World Problems	4, 5, 8, 11, 13, 36, 49, 68, 93, 101, 112, 136, 137, 163, 173	20, 22, 32, 52, 56, 89, 114, 116, 165, 182	17, 50, 58, 162	6, 7, 14, 18, 24, 26, 57, 66, 67, 71, 80, 82, 84, 103, 104, 108, 109, 110, 111, 113, 115, 122, 124, 126, 152, 179, 183, 184, 185, 187, 204	25, 73, 77, 117, 123, 125, 127, 132, 133, 150, 151, 157, 212, 231, 233, 236, 260
Shapes	17, 19, 41, 47, 62, 63, 64, 75, 87, 91, 92, 95, 99, 121, 131, 135, 139, 167, 181, 191, 195, 200, 201, 209, 219, 223, 235, 245, 249, 253, 257	20, 22, 96, 100, 192	16, 23, 33, 38, 39, 42, 43, 44, 45, 48, 50, 53, 58, 88, 119, 124, 149, 156, 162, 176, 196, 211, 216, 225, 227, 237, 238, 239, 250, 254, 259, 261, 263	51, 170, 179, 204, 230	25
Subtraction	60, 61, 68, 81, 83, 85, 86, 90, 93, 94, 98, 101, 130, 138, 168, 169, 171, 172, 174, 175, 177, 180, 186, 194, 195, 197, 206, 220, 221, 222, 233, 234, 244, 248, 256, 262	114, 116, 173, 182		230, 262	
Symmetry	41, 62, 63, 64, 189, 201, 217, 249	96, 192	16, 17, 23, 43, 53, 88, 196, 211, 216, 225, 227, 237, 263		
Temperature				226, 242, 243, 247, 258	
Time		193, 251, 264		108, 109, 110, 111, 113, 115, 193, 263	212
Weight	124, 126			124, 126, 179, 184	
Whole Numbers	2, 3, 8, 9, 11, 15, 21, 27, 28, 29, 30, 31, 34, 35, 36, 37, 40, 46, 49, 54, 60, 61, 65, 70, 74, 81, 83, 85, 86, 90, 93, 94, 97, 98, 101, 105, 107, 112, 120, 128, 130, 134, 138, 140, 145, 146, 147, 153, 155, 158, 159, 161, 163, 164, 166, 168, 169, 171, 172, 180, 188, 189, 190, 194, 200	52, 56, 59, 69, 72, 76, 79, 89, 102, 106, 114, 116, 118, 154, 160, 165		14, 18, 24, 57, 66, 67, 71, 76, 80, 82, 84, 103, 104, 108	117, 123, 125, 150, 151, 157

How to Use this Book

The skills and concepts presented spiral throughout this book. That means that you will see a topic dealt with for a few pages and then there will be a gap before it is covered again. We do that so the child has some time to develop and mature before dealing with more complex aspects of the concept.

Our suggestion is that you proceed through the book page by page. However, if your child is interested in a given topic and seems to want more, it would not be unreasonable for you to skip to the next level of that topic and do more of it.

Important: Keep learning fun and avoid frustrating the child. Work around the child's attention span. As an adult, you have a great advantage because most young children love to spend time with "big" people. If you keep learning fun, you will have an energetic pupil who looks forward to each lesson.

Throughout this book we have tried to use accurate vocabulary and notation. For example:

- *Number* refers to how many things are in a group.
- *Numeral* is used to refer to the written version of a number.
- *Number word* (e.g., two) is used when a number is spelled out.
- *Line segments* have a definite beginning point and end point.
- *Rays* have a definite beginning point and go on endlessly from there.
- *Lines* have no definite beginning point or end point but go on endlessly in both directions.

Regretfully, common usage of some mathematical terms has clouded the picture to the point that they are often used incorrectly. For example: *number* is often used in place of *numeral*, and *line* is often used when *line segment* should be used. We do bow to common usage with our use of "line of symmetry" where a line segment is used to bisect a figure.

There is a four-stage learning sequence that commonly helps students understand addition and subtraction. The stages are:

concrete – manipulation of objects
semi-concrete – pictures of objects
semi-abstract – tally marks to represent the objects
abstract – numerals

While each child progresses through the stages at different rates, they all go through these stages. Please note that this book does not include activities in the concrete stage. We assume that the student has completed the concrete stage through life and educational experiences (e.g., adding and subtracting cookies, blocks, etc.) and can proceed through the semi-concrete, semi-abstract, and abstract stages provided in this book. However, if at any time a child demonstrates difficulty (lack of understanding) conceptualizing any of the math concepts introduced in this book, we suggest you revert back to the concrete (manipulation of objects) stage. You can easily do this by recreating activities and lessons in this book using tangible objects.

A child with a thorough understanding of the math concepts taught in this book should be able to demonstrate an understanding of addition in the variety of settings provided in this book—not just one or two settings. **Moving to abstract levels too quickly may result in the child being able to memorize what is expected, but that is often accompanied by a lack of understanding.** Thus, we encourage you to provide sufficient exposure at each level before progressing to the next one. If a child struggles at a level, it is wise to return to the previous one and supply additional exposure and practice. This may seem time consuming, but in the long run tremendous time will be saved and understanding of the concept will be accomplished.

Addition

Facts are the first addition encounter. These are the items that must be memorized eventually. A fact is any digit (0, 1, 2, 3, 4, 5, 6, 7, 8, 9) added to any other digit (0, 1, 2, 3, 4, 5, 6, 7, 8, 9). These exercises, when written, should be given in both horizontal and vertical format (even at the concrete level). Abstractly, they should be shown as:

$$\begin{array}{r} 3 \\ \underline{+\ 4} \end{array} \quad \text{or} \quad 3 + 4 = \_\_\_ .$$

Terminology should be established as soon as possible. This will be helpful later when subtraction is encountered, as it will help link the two operations.

$$\begin{array}{rl} 3 & \textbf{addend} \\ \underline{+\ 4} & \textbf{addend} \\ 7 & \textbf{sum} \end{array}$$

As addition facts are learned, there are situations that can cause problems or help with the learning. Zero, for example, is difficult for many children to grasp, since it represents the absence of things. This idea of nothing being there is abstract to a child who is accustomed to talking about things that are possessed. As a result of this developmental factor, it is sometimes advisable to delay addition facts involving zero until the child has had a chance to mature a bit.

On the helpful side of the discussion, if a child learns fact pairs like 3 + 4 and 4 + 3, the memorization time can be decreased. Once a child realizes that reversing the order of the addends has no impact on the sum, far fewer facts need to be memorized because learning one brings the other with it. Eventually this reversal is formalized as the commutative property of addition, first on the set of digits, then later on the sets of counting numbers, whole numbers, integers, rational numbers, and, finally, real numbers. This is a good example of how the groundwork laid in first grade carries through all of the years of education as an idea is visited again and again.

Often sums less than 10 are dealt with exclusively at first. This helps the child grasp important concepts without having to deal with place value (numbers involving two or more digits, like 23, 456, and so on). If a child has mastered sums less than 10 and is struggling with place value, challenge can be provided with exercises involving three addends (numbers to be added) with a sum that is less than 10 (2 + 3 + 4 = ?).

Once place value is introduced, the child can deal with multiples of 10 (10, 20, 30, 40, ...) plus multiples of 10 where the sums are less than 100. (Before sums of 100 or more are encountered, additional place value discussion would be needed.) Sums of multiples of 10 can seem easy to a child who has a solid grasp on the addition facts. Something like 20 plus 30 would be approached as "2 tens" plus "3 tens". This sounds much like "2 apples" plus "3 apples", only, here, the "tens" has the added significance related to place value. Still, the sum is 5 "tens", in this case with the emphasis being on "2 + 3" AND "tens". The

"2 + 3" is old familiar territory (if the child has not memorized the facts, it is difficult to justify going on) and the only new thing is the "tens". Soon the child realizes that 20 + 30 is really just like adding 2 and 3 except that there are zeros at the end of both the 2 and 3, and, therefore, there has to be a zero at the end of the 5, too.

Once addition of multiples of 10 is mastered, addition involving 2-digit addends, with no regrouping, can be tackled. This could be either 2-digit plus 2-digit addends (24 + 65), or 2-digit plus 1-digit (52 + 6). Exercises involving 1-digit plus 2-digit addends should not be overlooked. Even though this reversal of order might seem like a small change, and even though most books show the problems as 2-digit plus 1-digit, both orders are important for conceptualization. Once conceptualization of addition is achieved, things like size and order of size of addends will not matter. However, here at the formative stages, these seemingly small issues are significant.

There is not agreement on whether 2-digit plus 2-digit should come before or after 2-digit plus 1-digit addition. Generally, it is agreed that one should follow the other. Considering 2-digit plus 2-digit, the child encounters something like 12 + 45. Initially, base 10 blocks could be used to show the two addends, using a ten and 2 units to be combined with 4 tens and 5 units. That could be followed with pictures of the base 10 blocks and then statements like 1 ten and 2 ones + 4 tens and 5 ones. After these stages, abstract representation can be encountered and the first one is a natural extension to the idea of 1 ten and 2 ones + 4 tens and 5 ones. It would be:

10 + 2
+ 40 + 5. This will appear to be pretty easy for a child who has mastered addition facts and the idea of adding multiples of 10, because they have already done exercises that look like each part of the problem. That is, they will have done problems like:

10 **2**
+ 40 and **+ 5**.

Rather soon, the setting can be expressed as:

12
+ 45. (The color is added to help connect the prior step in the sequence and would be eliminated as soon as possible.)

A similar developmental sequence would be used for problems involving 2-digit plus 1-digit addends (and the reversed order). The concrete, semi-concrete, and semi-abstract stages of learning are critical at this point and should not be overlooked or rushed through. They provide an essential ingredient that leads to conceptualization of addition.

Regrouping in addition will be the next step and would involve both 2-digit + 2-digit and 2-digit + 1-digit examples. Again, the two problem types should be kept separate at first and in the case of 2-digit + 2-digit addends, regrouping should be limited to coming out of the ones place only (24 + 37 and not 42 + 73), since regrouping out of the tens involves hundreds, which involves another discussion of place value first. Besides, one step at a time saves time as the "big picture" is created. Typically, regrouping in addition is not encountered in first grade.

Subtraction

Development of subtraction is usually several months behind that of addition, but, even still, the cycles and operations will overlap. Confusion is possible, which only emphasizes the need to do each problem type concretely first and then progress through the stages to abstraction.

Facts are the first subtraction encounter. As with addition facts, these too must be memorized. Terminology becomes an issue here. Different terms have been and continue to be used for subtraction situations. However, the most beneficial terminology is:

$$\begin{array}{rl} 7 & \textbf{sum} \\ \underline{-\,3} & \textbf{addend} \\ 4 & \textbf{missing addend.} \end{array}$$

Notice how these are the same words that were used with addition except for "missing". Think back to how you check subtraction. More than likely, you add the answer (4) to the number that was subtracted (3) to get the number subtracted from (7). There you have "addend" plus "addend" equals "sum", just like in addition. These word links can prove helpful for children, particularly those who struggle.

Initially the subtraction facts could be limited to a sum that is less than ten, although that may not be as necessary as it was with addition, since place value should have been encountered with addition already. However, particularly for the child who struggles with subtraction, avoiding the place value issue for a while could be helpful.

All of the sequencing and developmental discussion given with addition would now be repeated for subtraction. It is unlikely that subtraction will be as fully developed as addition by the end of the child's first grade experience.

Virtual Manipulatives Website

We suggest parents and teachers using this product explore the National Library of Virtual Manipulatives at http://nlvm.usu.edu/en/nav/index.html. Click on "Virtual Library" at the top of the page. Then click on the Number and Operations row and the Pre K-2 column. If you click on the second entry, "Base Blocks," you can create your own problem. To see how it works, click on the unit cube and drag it to the space below it. Do that at least 12 times. Then, using your mouse, draw a rectangle around 10 of the pieces you see in the workspace. You will see those 10 units trade for one ten. That is a powerful image for a child.

Please note that there are base 10 blocks for addition and subtraction on that same Number and Operations row and Pre K-2 cell on the table. It would be worthwhile for you to browse the other activities on that page. For example:

- Algebra and K-2 category, "Complete the Pattern"
- Geometry and K-2 category, "Tangram Puzzles"
- Measurement and K-2 category, "Attribute Trains"
- Data Analysis and Probability, "Bar Chart"

All of these virtual activities can be used for additional practice and to stimulate creative notions in each child.

The activities in this book are written to the standards of the National Council of Teachers of Mathematics.

Count the number of stars in each set. Then draw a line segment between that set and the numeral that goes with it.

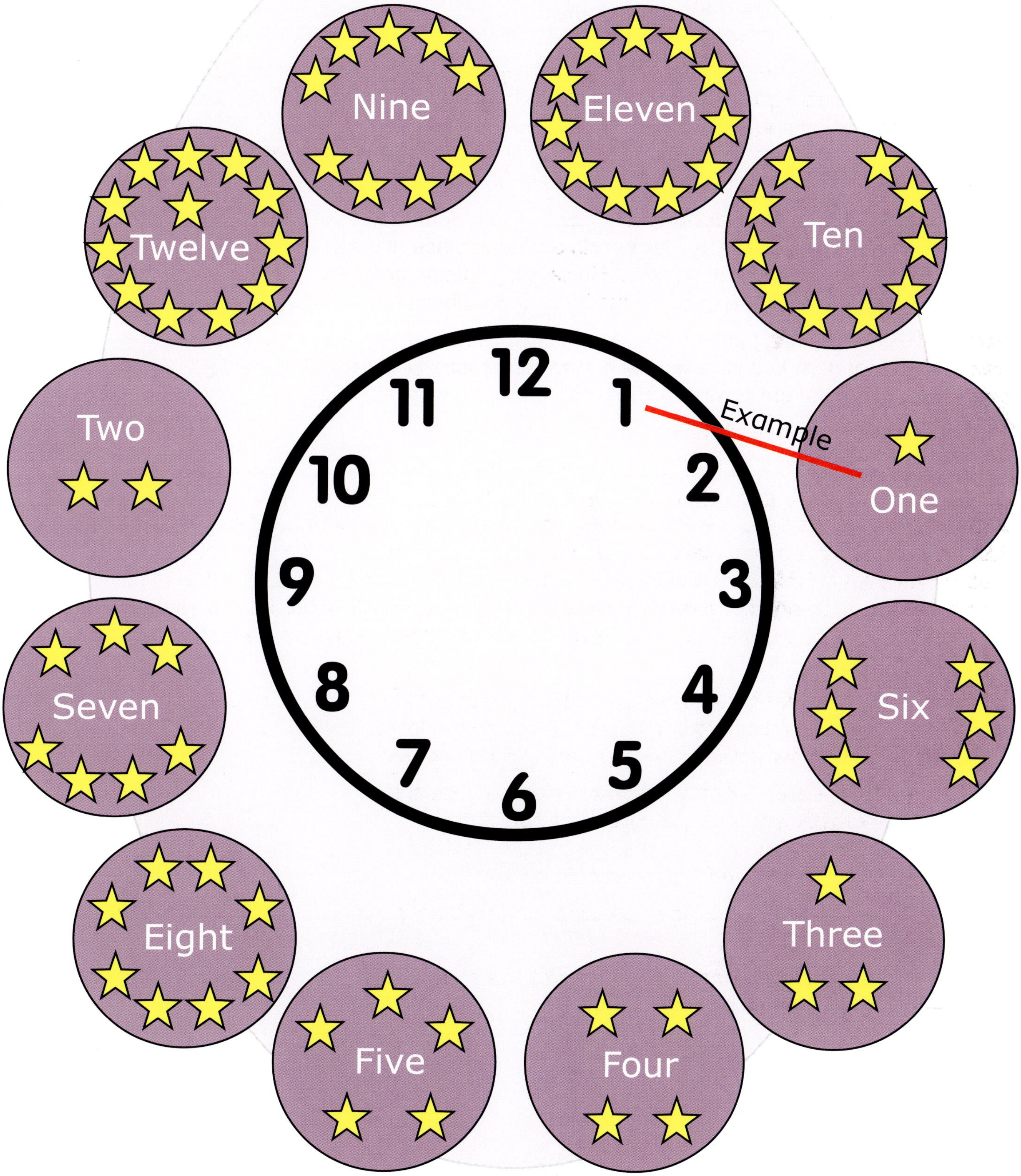

Trace the numerals, then write the numerals that continue the pattern.

1 ____ 3 ____ ____ ____ 7 8 9

2 4 6 ____ 10 ____ 14 ____

1 3 5 ____ 9 ____ 13 ____

15 ____ 13 ____ 11 ____ ____

1 ____ 3 ____ 5 ____ 7 ____ 9 ____

Numeral Guide 0 1 2 3 4 5 6 7 8 9

Trace each numeral, then draw a line segment to the matching number word.

Example

one	4	eleven	14
two	3	twelve	11
three	1	thirteen	12
four	2	fourteen	20
five	10	fifteen	13
six	9	sixteen	18
seven	7	seventeen	17
eight	8	eighteen	15
nine	6	nineteen	16
ten	5	twenty	19

Add the number of blocks in each set. Then write and say each number sentence.

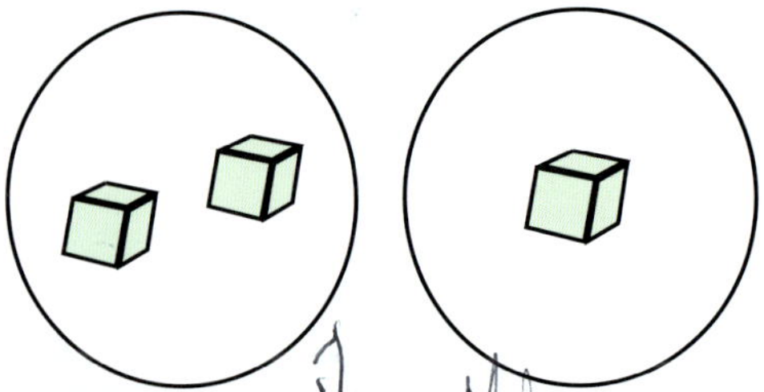

_____ + _____ = _____

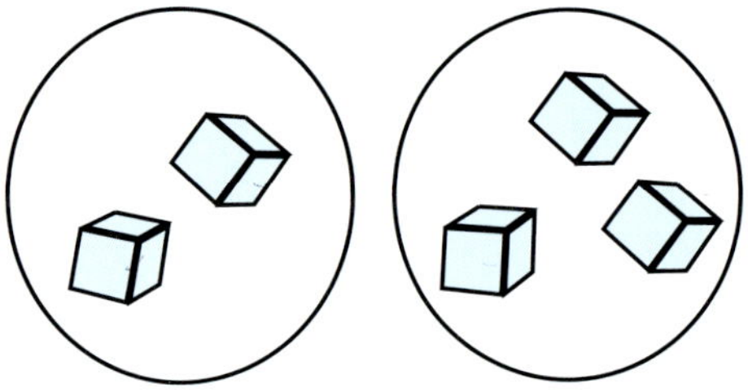

_____ + _____ = _____

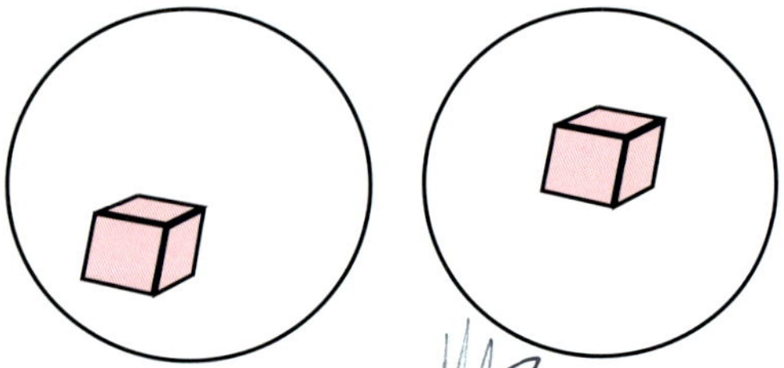

_____ + _____ = _____

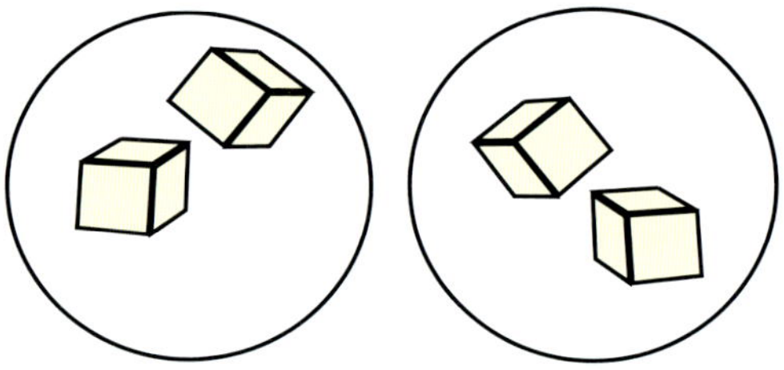

_____ + _____ = _____

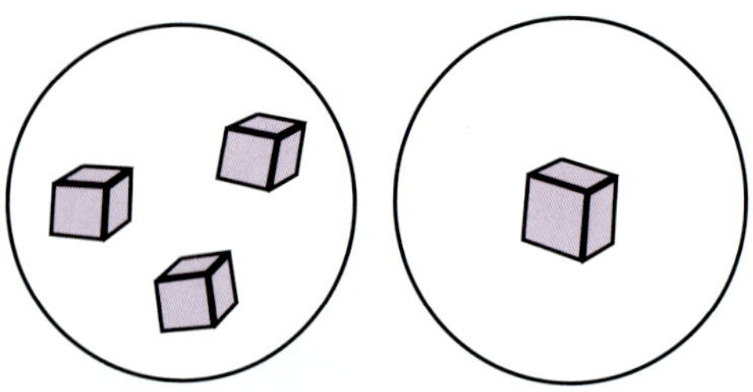

_____ + _____ = _____

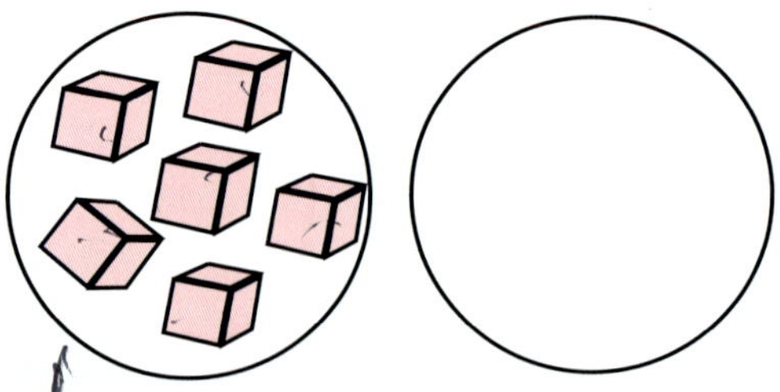

_____ + _____ = _____

Numeral Guide 0 1 2 3 4 5 6 7 8 9

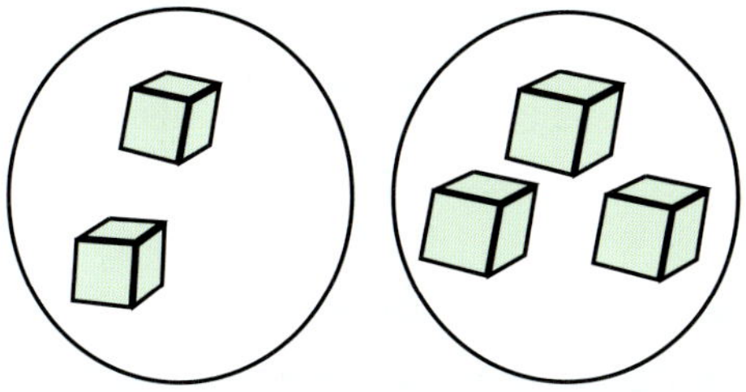

____ + ____ = ____

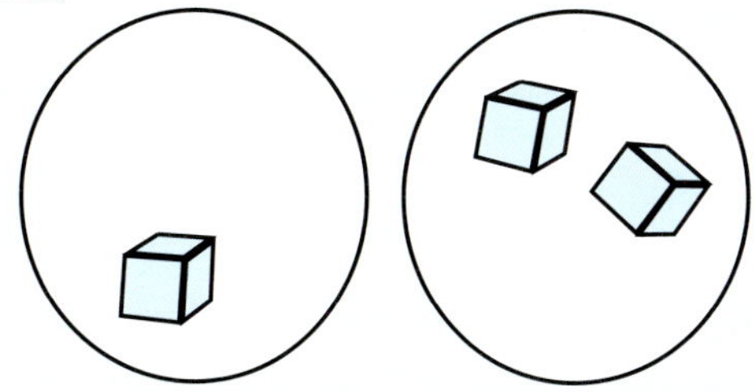

____ + ____ = ____

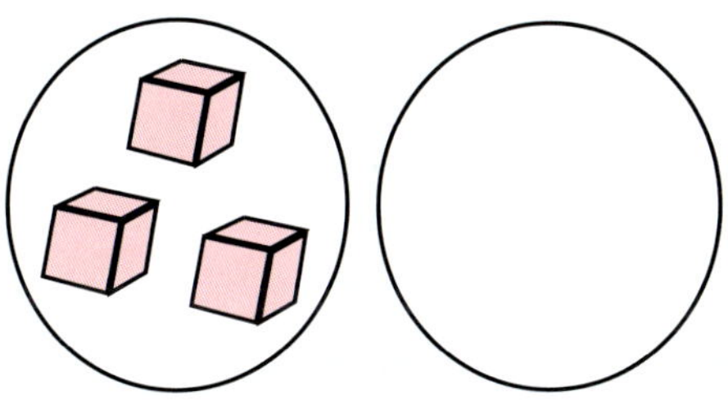

____ + ____ = ____

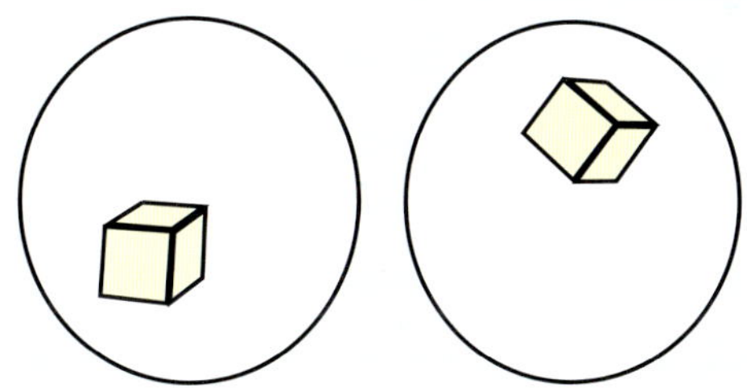

____ + ____ = ____

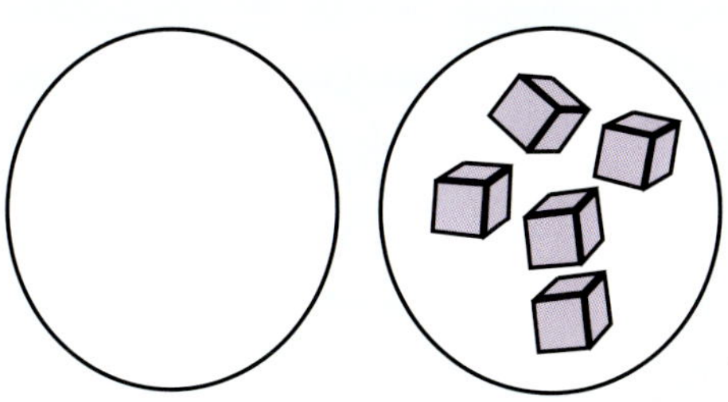

____ + ____ = ____

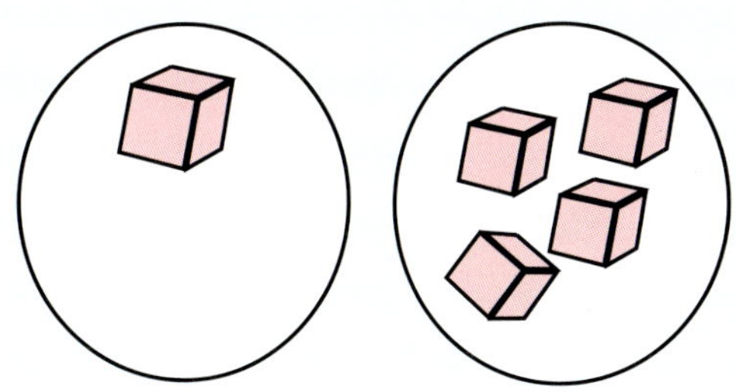

____ + ____ = ____

Numeral Guide 0 1 2 3 4 5 6 7 8 9

Write each numeral, then say each number sentence.

A penny is worth 1 cent (1¢).

A nickel is worth 5 cents (5¢).

A dime is worth 10 cents (10¢).

5¢ = 1¢ + ____ + ____ + ____ + ____

1¢ + ____ + ____ + ____ + ____

10¢ =

\+ ____ + ____ + ____ + ____ + ____

10¢ = 5¢ + ____

10¢ = ____ + ____ + ____ + ____ + ____ + ____

Cross out the extra coins.

Add the number of balls in each set. Then write and say each number sentence.

____ + ____ = ____

____ + ____ = ____

____ + ____ = ____

____ + ____ = ____

____ + ____ = ____

____ + ____ = ____

Numeral Guide 0 1 2 3 4 5 6 7 8 9

Trace the numerals, then write the numerals that continue the pattern.

1 ____ 3 4 ____ 6 ____ 8 ____ 10

11 ____ ____ 14 ____ 16

1 2 ____ ____ 5 ____ 7 ____ 9 ____

11 ____ 13 ____ 15 ____

6 ____ ____ 9 ____ ____ 12

8 ____ 10 11 ____ ____ 14

Pairs

Partners

2 of the same thing equals 1 **pair** or a set called **partners**.

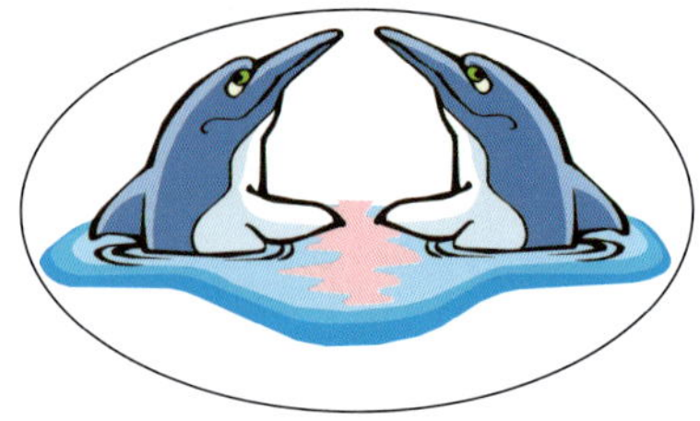

Count the cars.
Circle the pairs (partners).

How many pairs? ________

How many yellow cars (count by twos). ________

Count the pencils.
Circle the pairs (partners).

How many pairs? ________

How may red pencils? ________

Add the number of bears in each set.
Then write and say each number sentence.

7 + 2 = 9

5 + 5 = 10

4 + 5 = 9

5 + 3 = 8

1 + 7 = 8

4 + 6 = 10

Numeral Guide 0 1 2 3 4 5 6 7 8 9

Count the hearts on each card, then draw line segments matching the cards that equal 7.

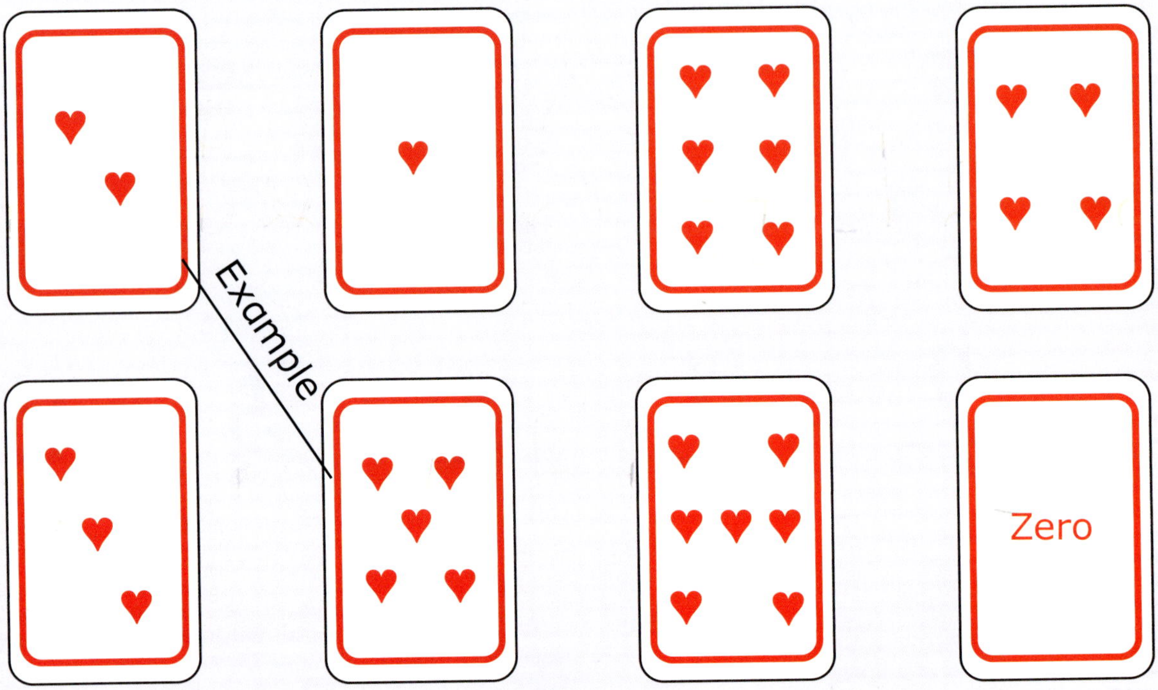

Count the clubs on each card, then draw line segments matching the cards that equal 8.

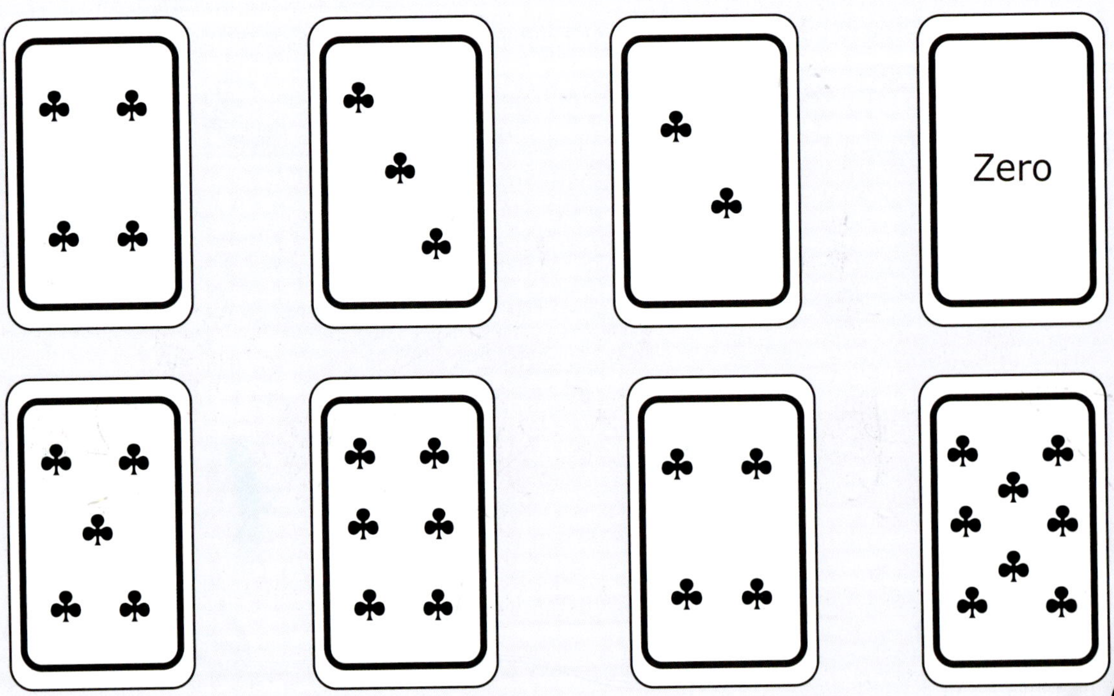

Skip and count by 5.

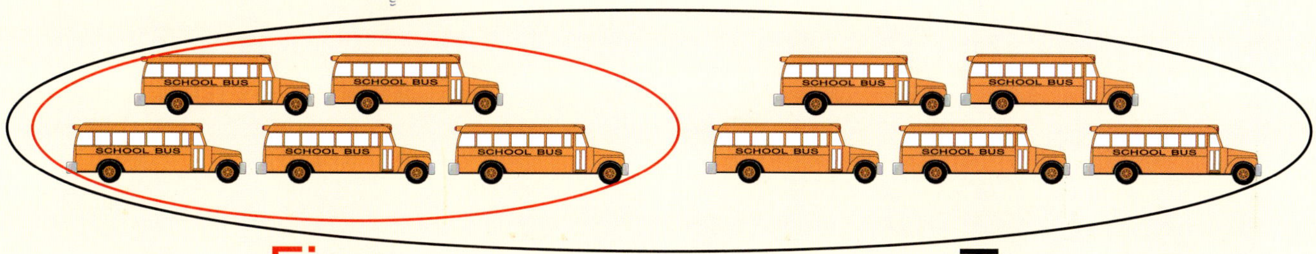

Five Ten

Circle sets of 5, then skip and count by 5.

How many pencils? _______

How many sets of 5? _______

Circle sets of 5, then skip and count by 5.

How many books? _______

How many sets of 5? _______

How many cents are in each set of coins?

How many cents? ______¢

How many cents? ______¢

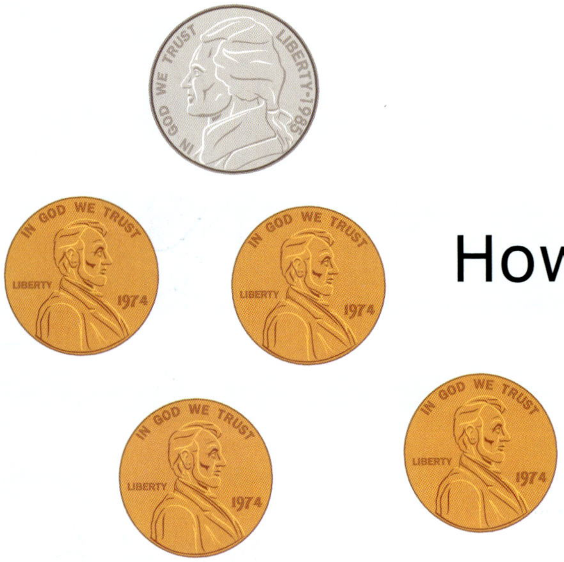

How many cents? ______¢

Trace the numerals, then match the numeral to the number sentence it completes.

0 + 0 =	4
1 + 1 =	10
2 + 2 =	0
3 + 3 =	12
4 + 4 =	2
5 + 5 =	14
6 + 6 =	6
7 + 7 =	18
8 + 8 =	8
9 + 9 =	16

When you divide something into two parts and both halves look exactly the same, we say the object is **symmetric**, or has **symmetry**.

For example, this square is divided by the red line segment into two parts that are exactly the same. The line that does the dividing is called the **line of symmetry**.

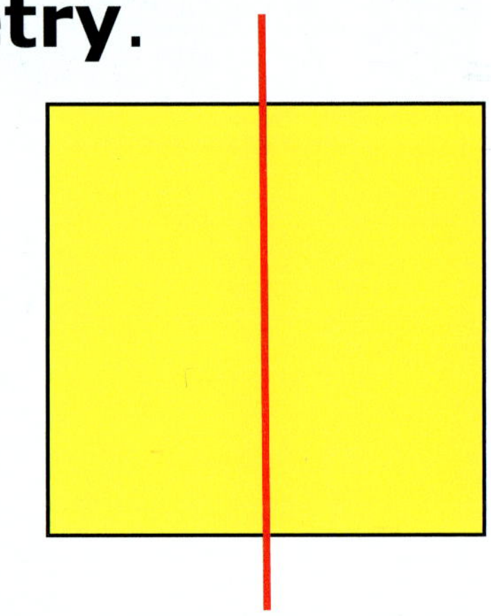

Circle all the symmetric objects below. Put an X over all the objects that are not divided by a line of symmetry.

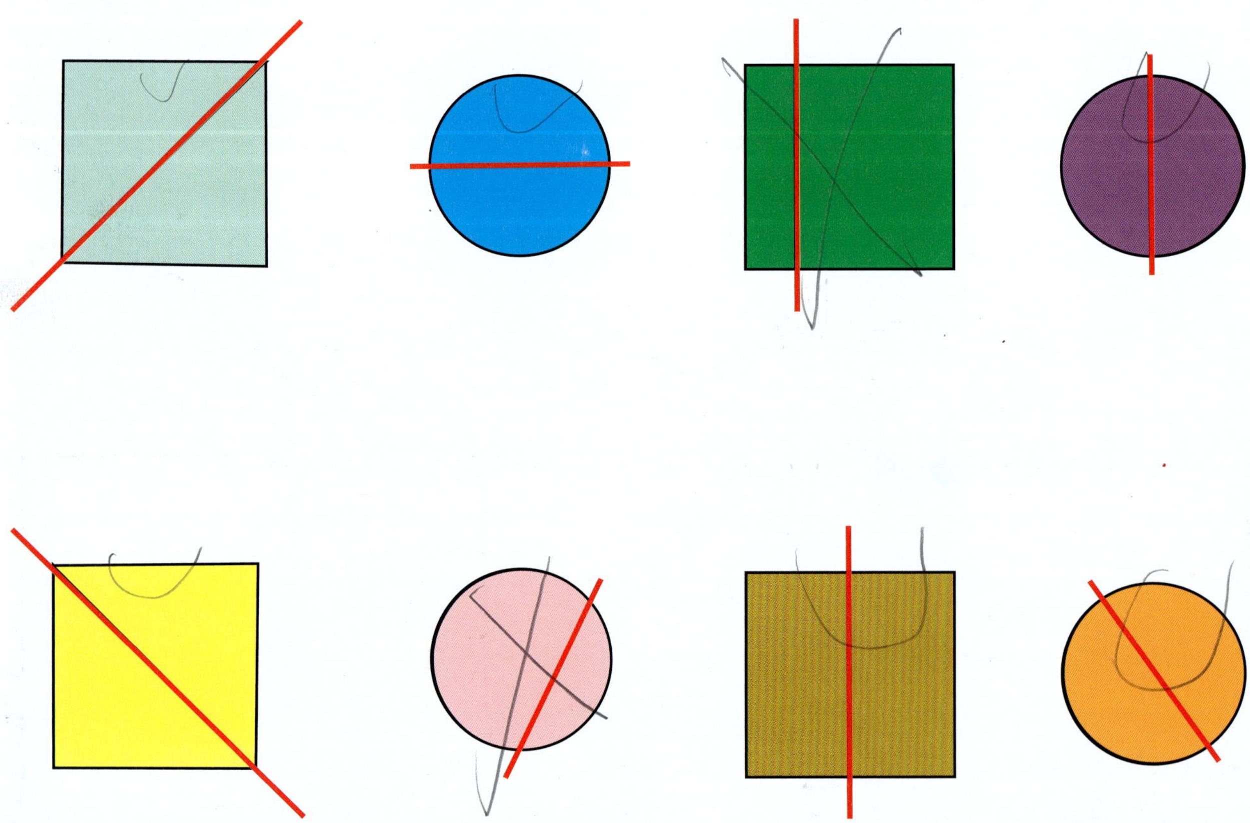

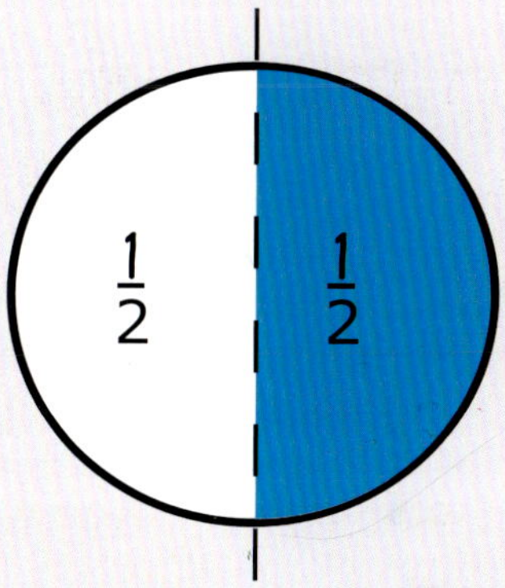

Two equal parts are called **halves**. One half of the circle is white and one half is blue. $\frac{1}{2}$ means 1 half.

Circle objects that show halves, then color one of the halves.

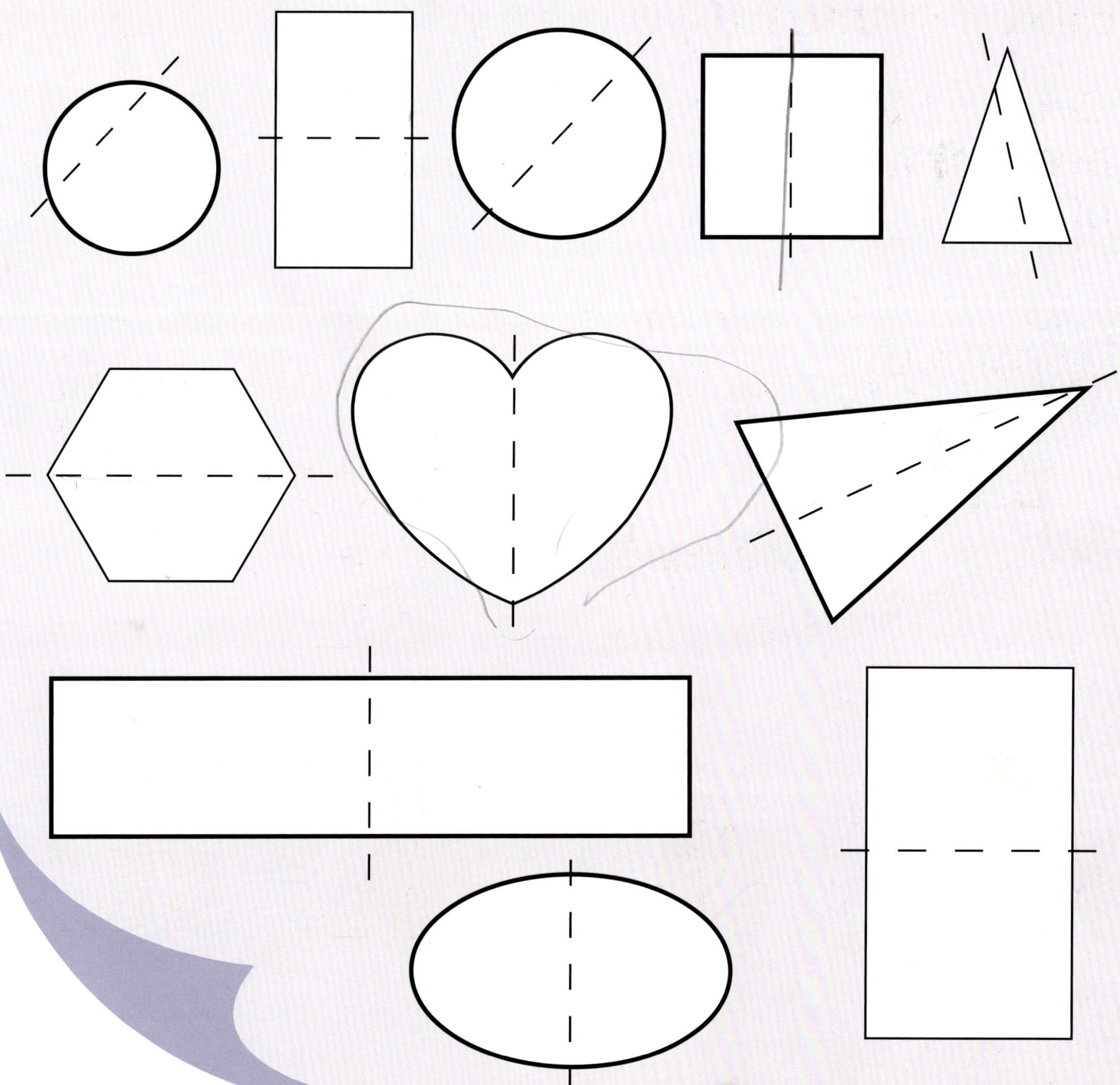

Touch and count each set, then write the sum of the cents.

+

+

= ______¢

1¢ 1¢ 1¢

+

+

= ______¢

5¢ 10¢ 5¢

+

+

= ______¢

5¢ 1¢ 1¢

Count the hearts on each card, then draw line segments matching each pair of cards that add up to 9.

Draw and color the next two shapes that would repeat the pattern in each row.

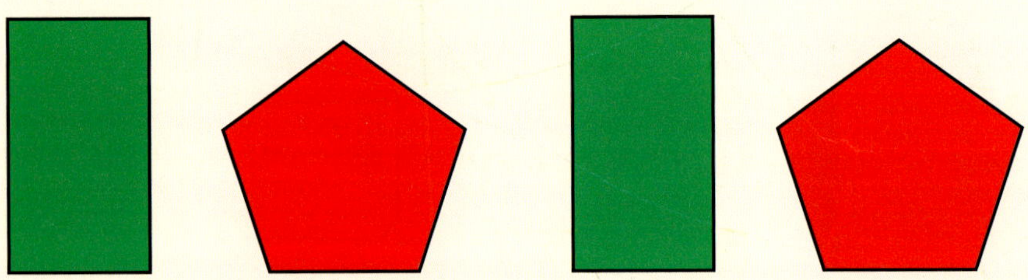

Trace each numeral, then draw line segments to the matching number word and quantity.

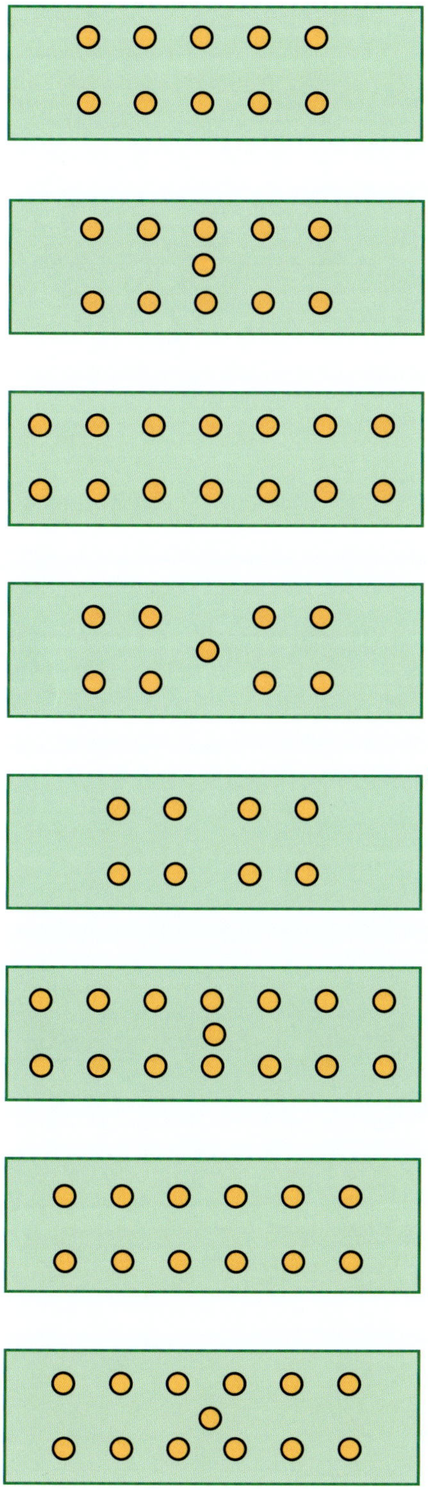

eight	10
nine	8
ten	9
eleven	14
twelve	15
thirteen	12
fourteen	11
fifteen	13

Example

Example

Draw and color the next two shapes that would repeat the pattern in each row.

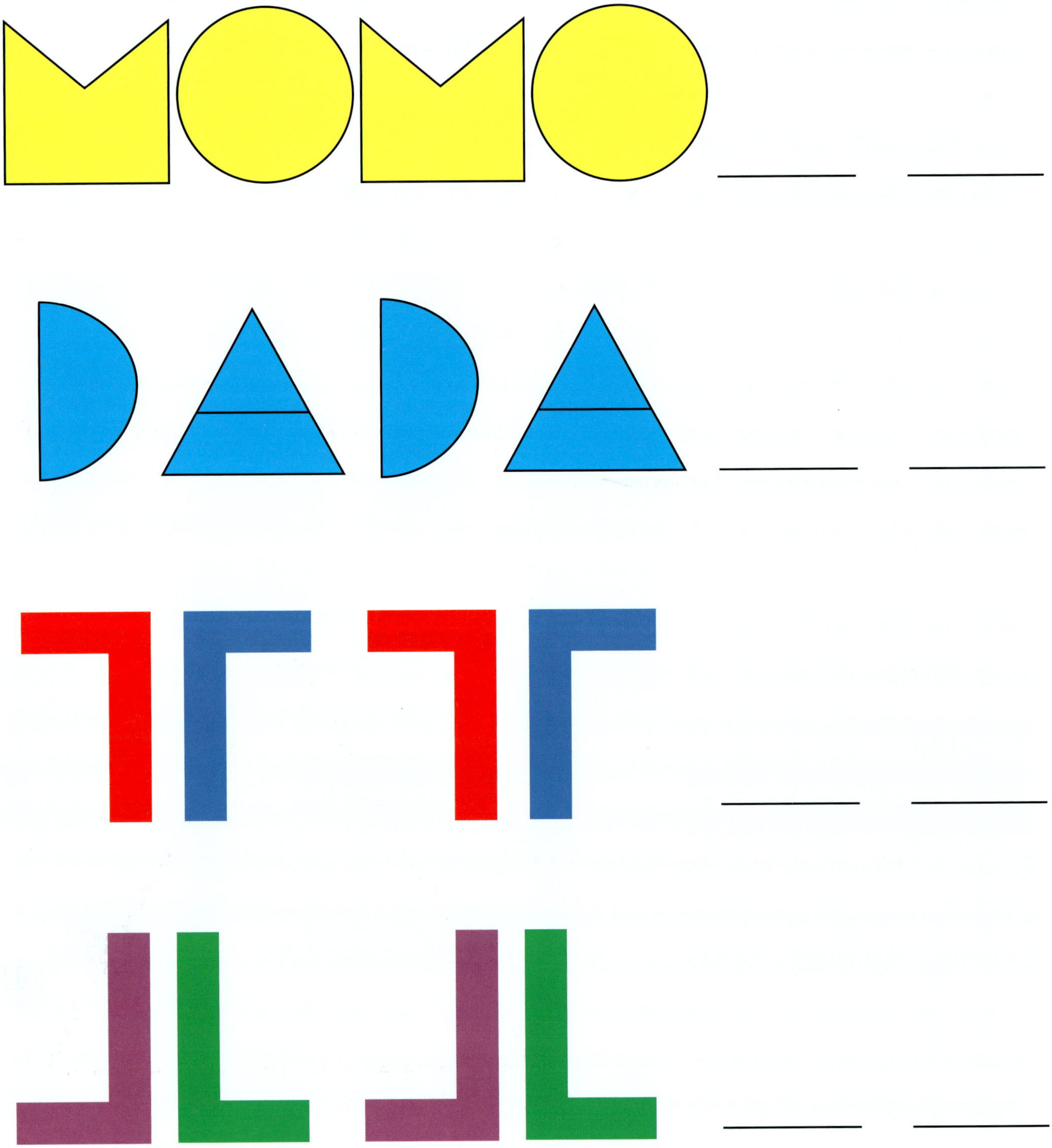

Draw a line of symmetry on each shape.

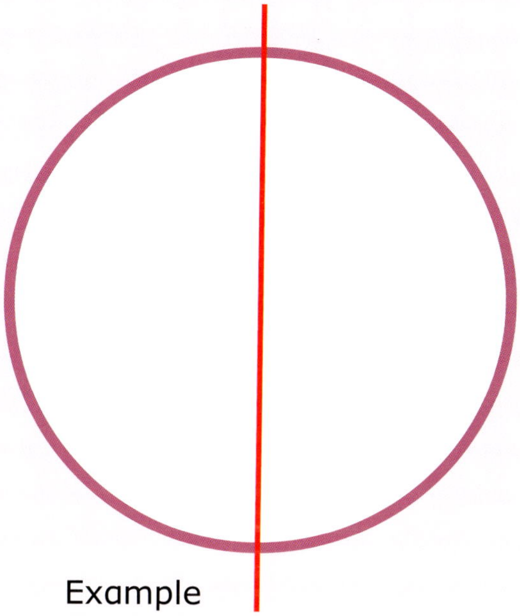

Example

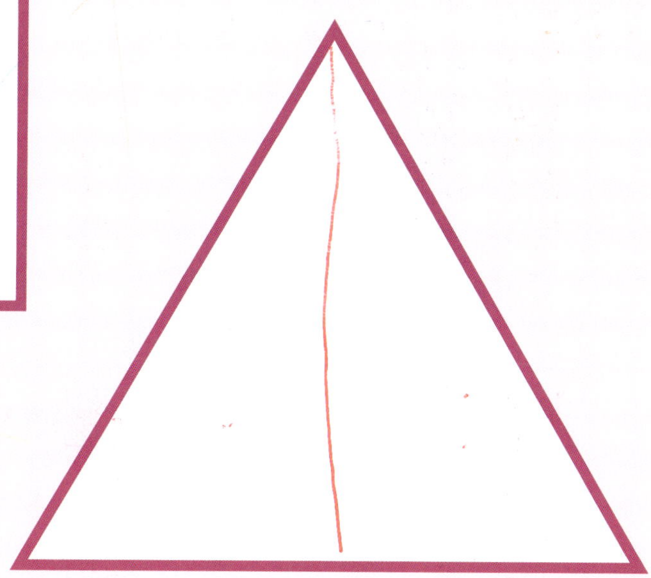

How many halves are on the page? __________

Draw line segments to connect the matching amounts of cents.

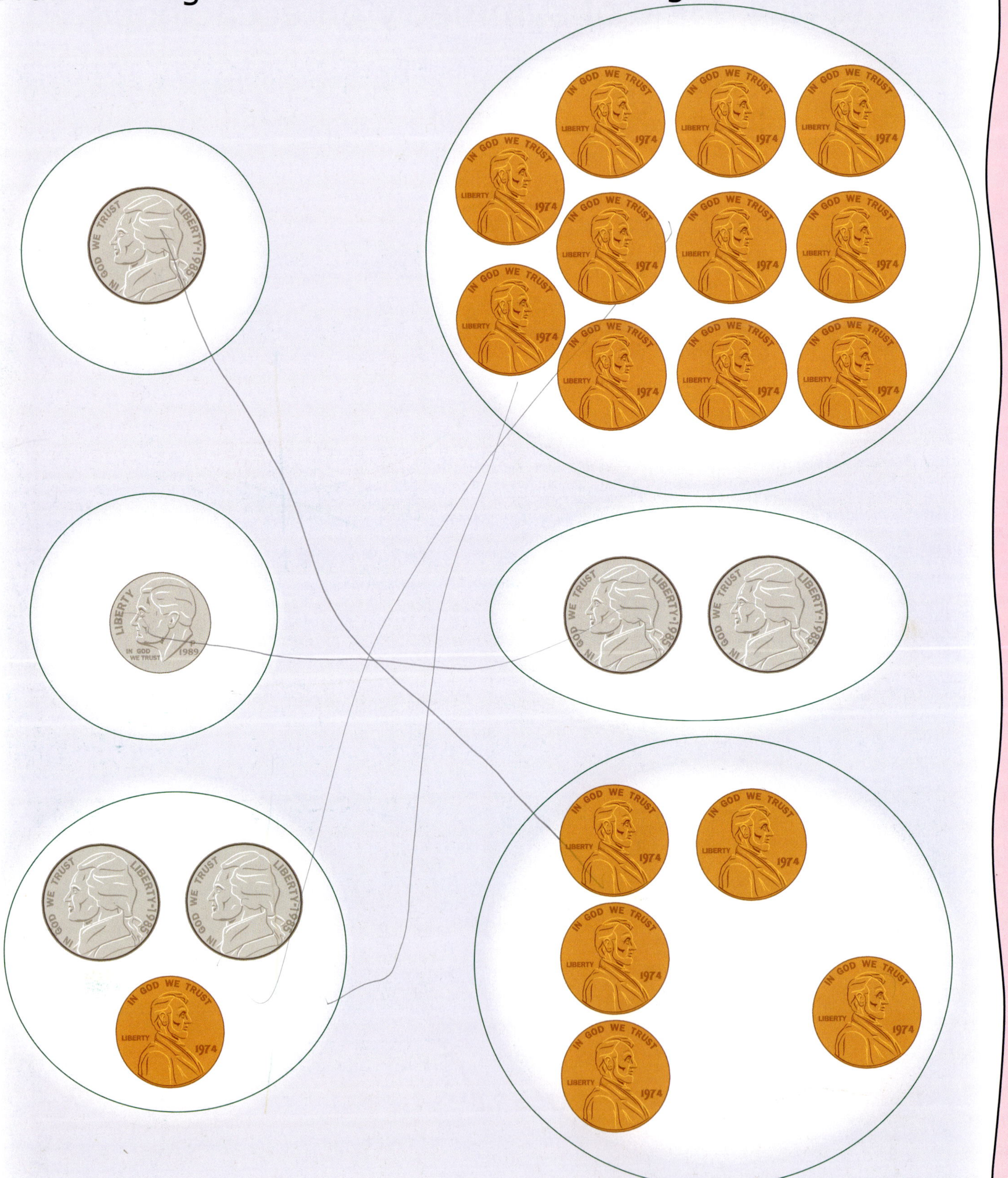

Numeral Guide 0 1 2 3 4 5 6 7 8 9

Put an X over the thing in each set that does not belong, then explain your answer.

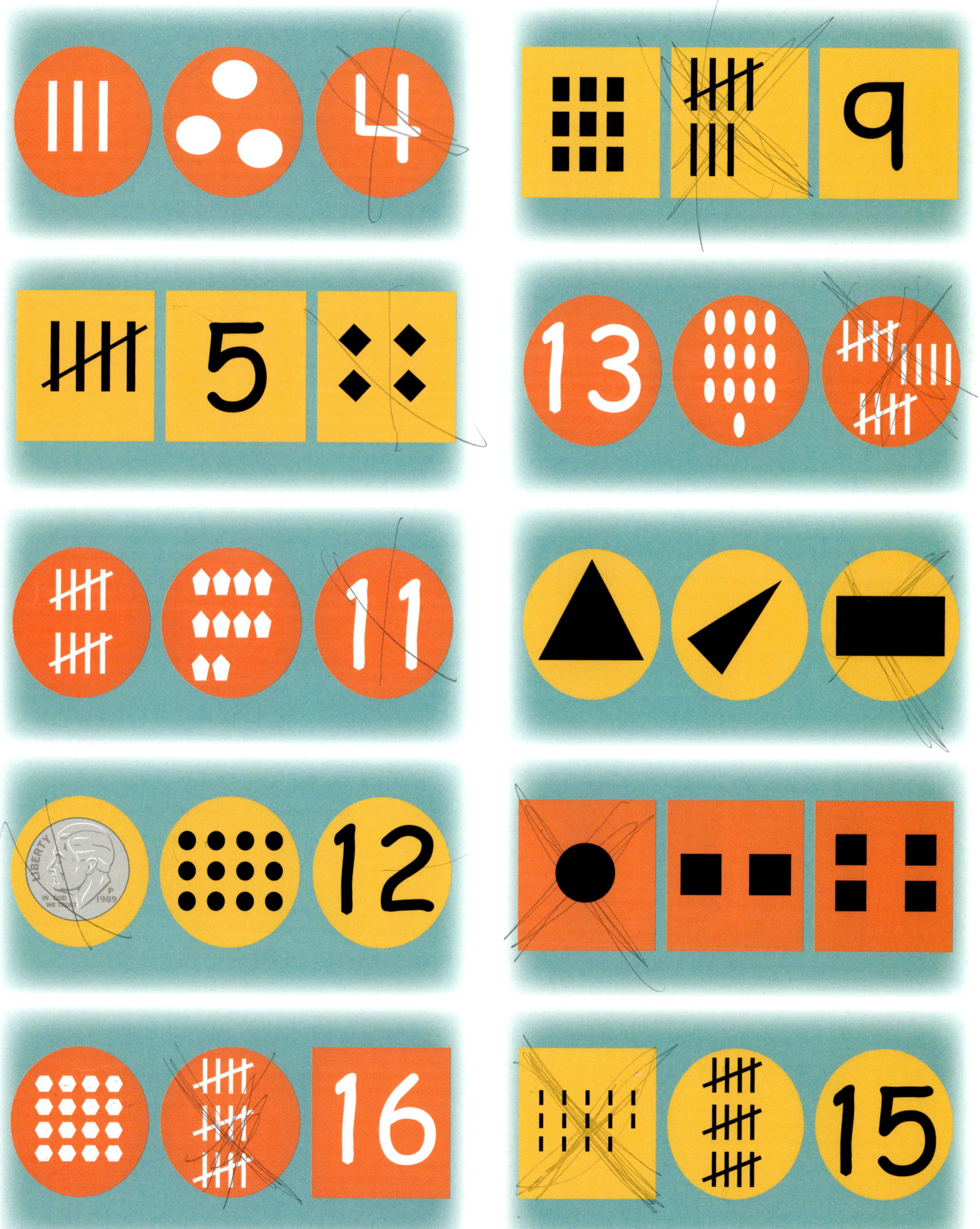

Write the number of cents in each set.
Then draw line segments to connect the sets with the same number of cents.

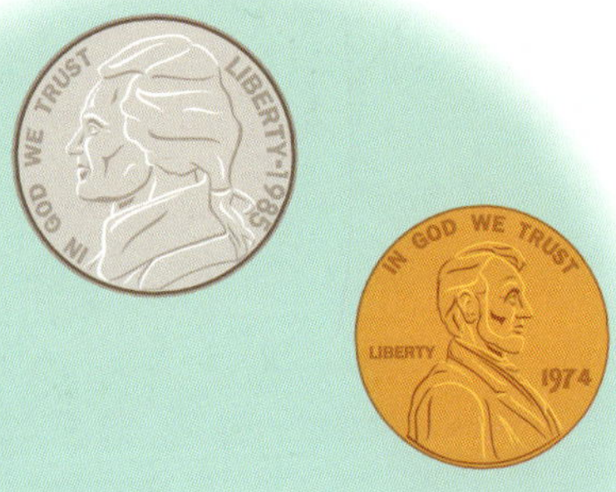

______¢

______¢

______¢

______¢

______¢

______¢

Numeral Guide 0 1 2 3 4 5 6 7 8 9

Count the tally marks in each set. Then write the sum of the tally marks and complete the number sentence.

|||| + ||| = ( )

4 + 3 = 7

||| + |||||| = ( )

____ + ____ = ____

|||||| + ||| = ( )

____ + ____ = ____

~~||||~~ + |||| = ( )

____ + ____ = ____

Use a number word from the choice box to make each sentence true. Words can be used more than once.

zero one two three four five six seven eight nine ten eleven twelve thirteen

One and __________ equals four.

Three and __________ equals four.

__________ and two equals four.

__________ and five equals five.

Five and __________ equals five.

Eight and two equals __________.

Ten and __________ equals twelve.

__________ and three equals eight.

__________ and seven equals fourteen.

Eight and __________ equals eleven.

Trace each numeral, then draw line segments to the matching number word and tally marks.

one	2
two	5
three	4
four	3
five	1
six	8
seven	10
eight	6
nine	12
ten	7
eleven	9
twelve	11

Example Example

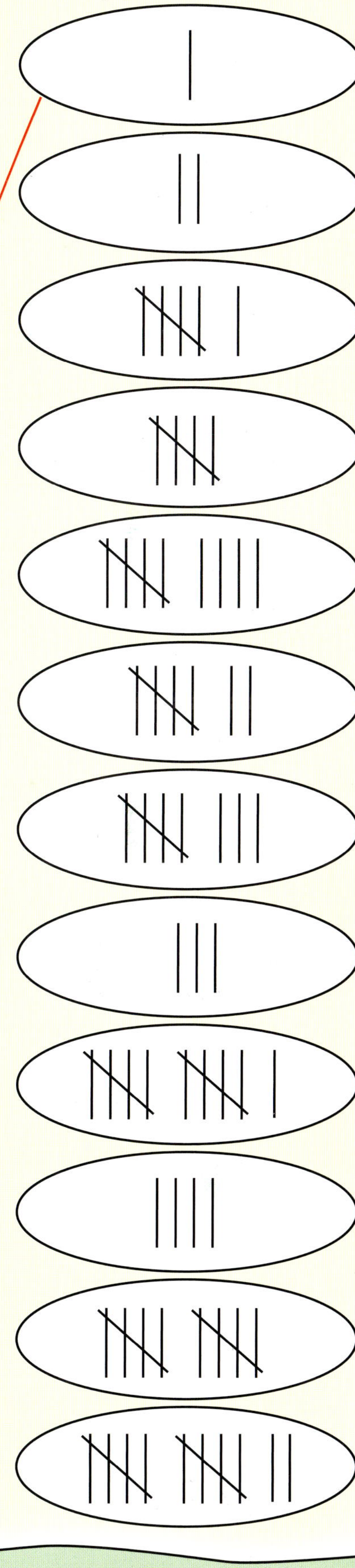

Count the tally marks in each set.
Then write the sum of the tally marks
and complete the number sentence.

________ + ________ = ________

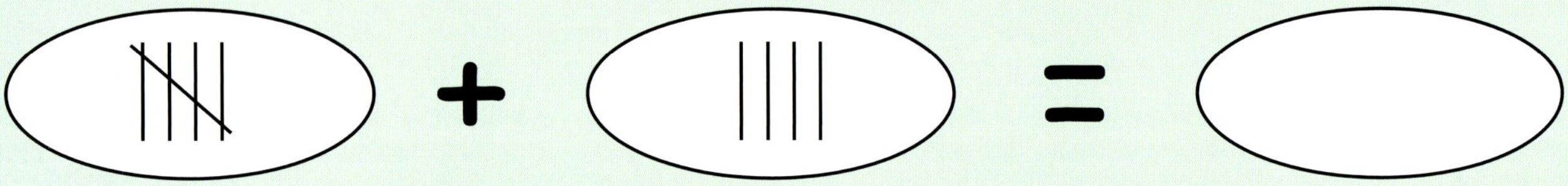

________ + ________ = ________

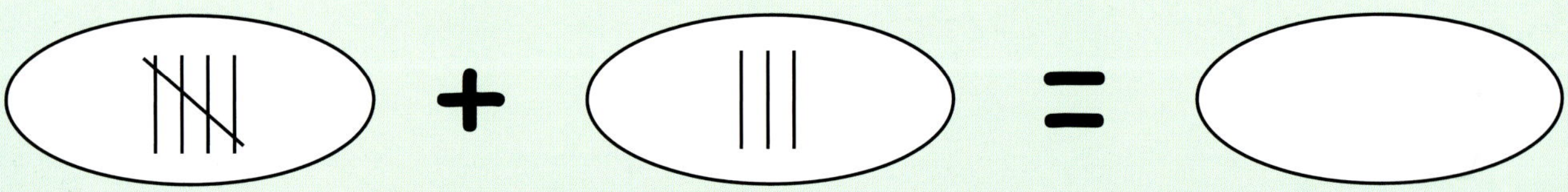

________ + ________ = ________

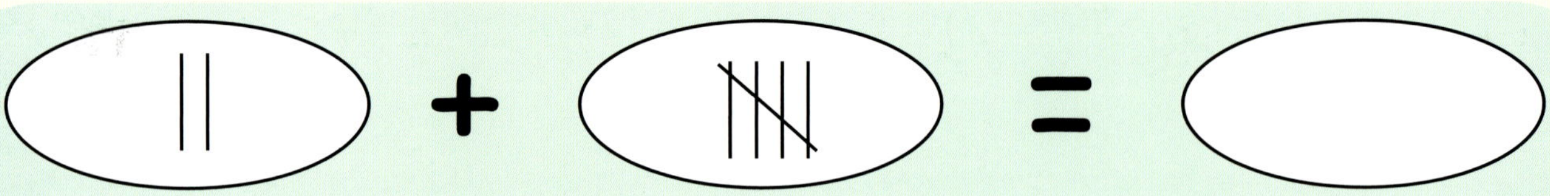

________ + ________ = ________

Read each number word, then complete each number sentence.

zero + one = one	zero + two = two
0 + 1 = ____	0 + 2 = ____
zero + three = three	zero + four = four
0 + ____ = 3	____ + 4 = 4
zero + five = five	zero + six = six
0 + 5 = ____	____ + 6 = 6
zero + seven = seven	zero + eight = eight
____ + 7 = 7	0 + ____ = 8
zero + nine = nine	zero + ten = ten
0 + 9 = ____	0 + ____ = 10

Numeral Guide 0 1 2 3 4 5 6 7 8 9

Do the actions, then count aloud by repeating each pattern five times.

clapping

spinning

laughing

jumping

Draw a square inside all of the triangles.

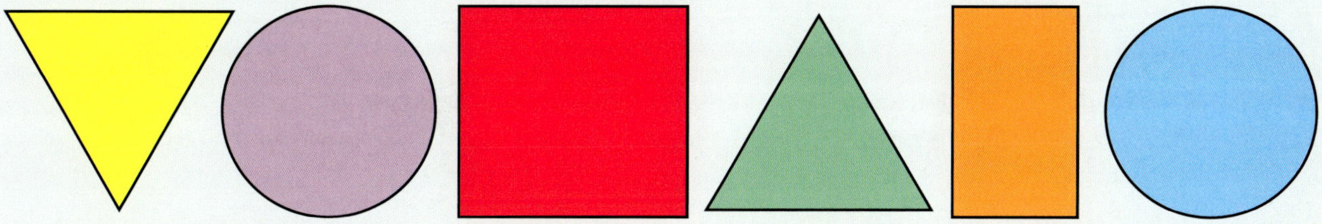

Draw a triangle inside all of the circles.

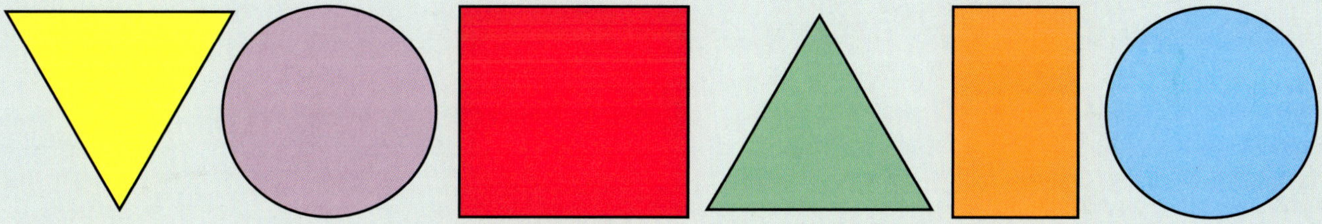

Draw a circle inside all of the rectangles.

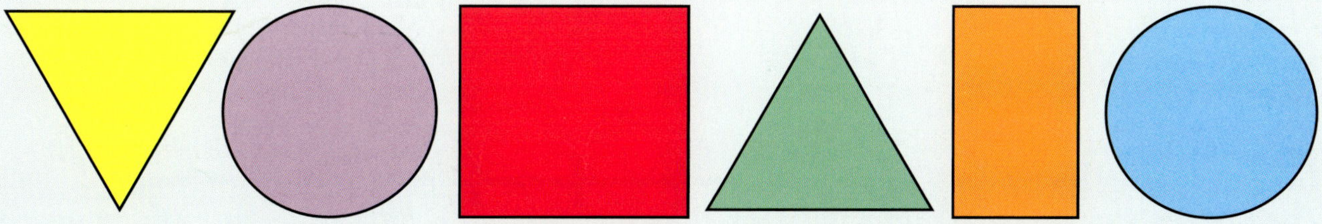

Draw a triangle inside all of the shapes that have corners.

Use the number line to show the numerals in each problem, then write the sum.

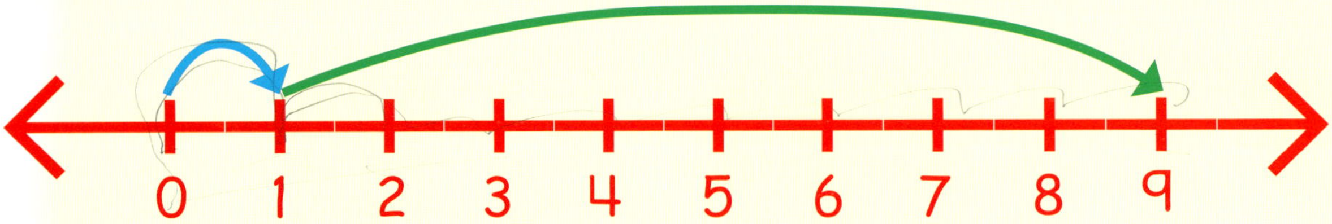

1 + 8 = ____

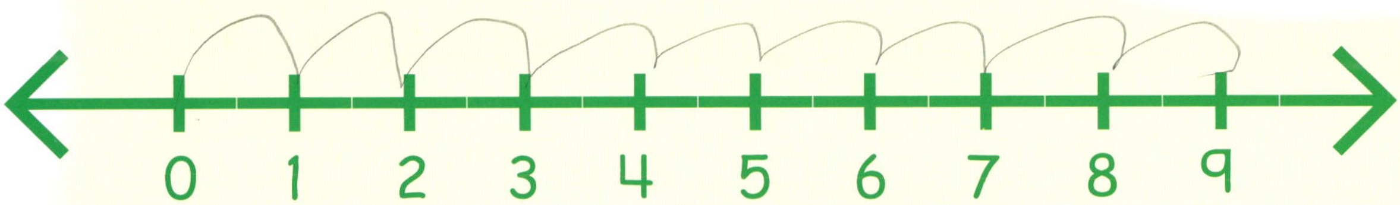

3 + 6 = ____

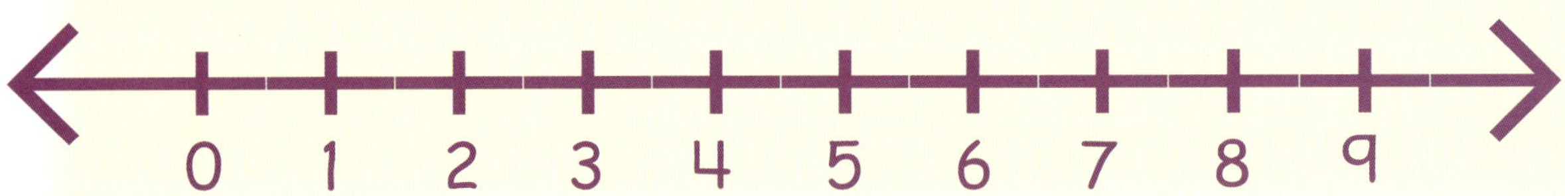

4 + 4 = ____

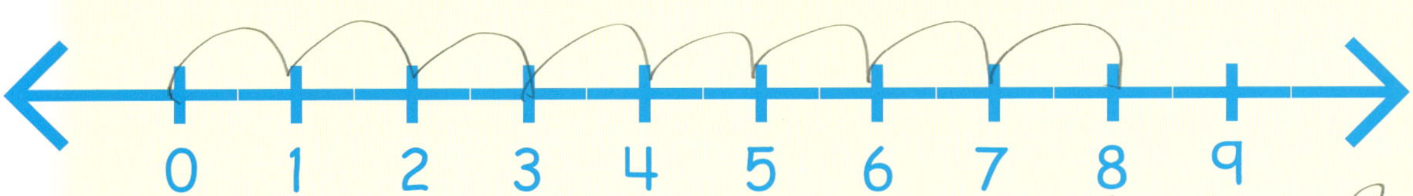

3 + 5 = ____

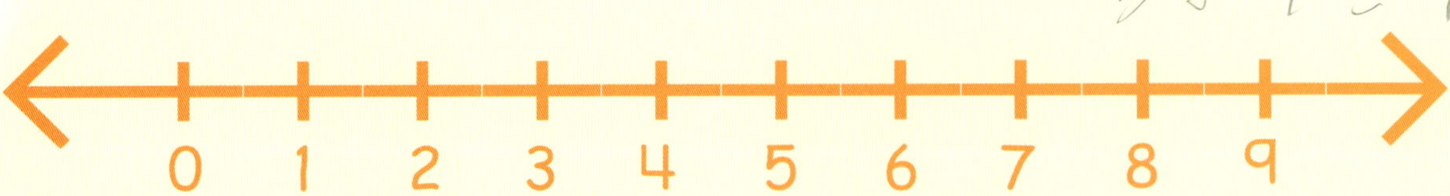

7 + 2 = ____

Use the number line to show the numerals in each problem, then write the sum.

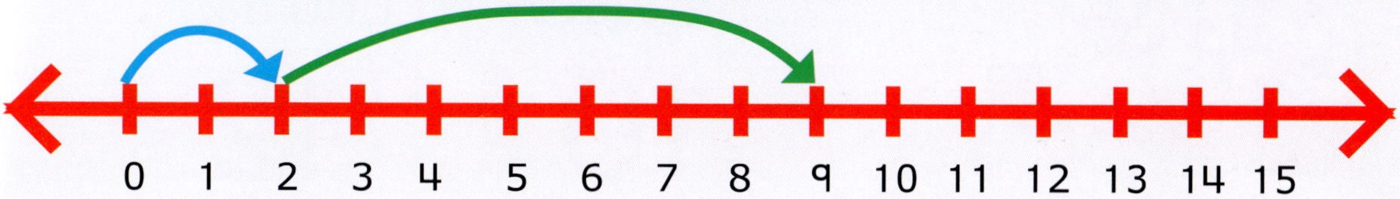

2 + 7 = ____

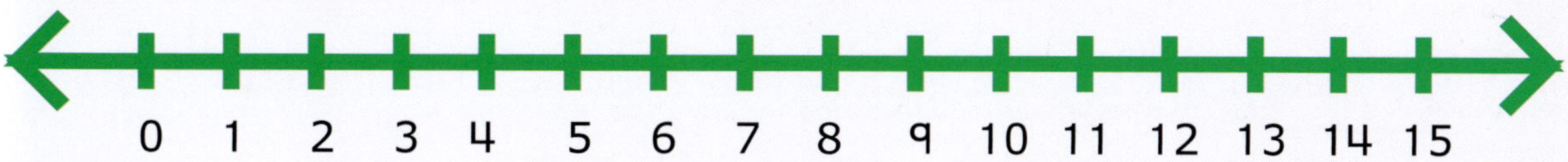

10 + 5 = ____

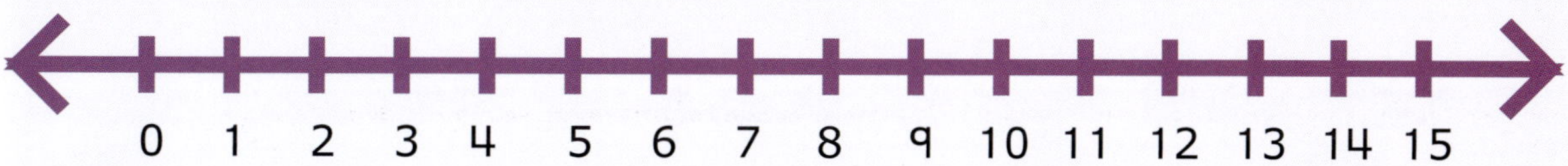

4 + 11 = ____

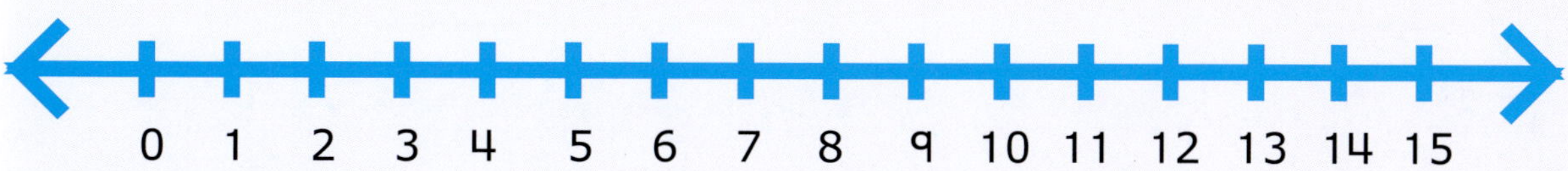

6 + 6 = ____

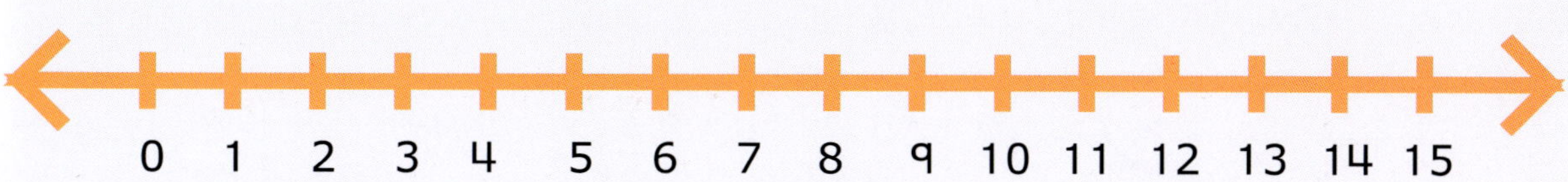

7 + 8 = ____

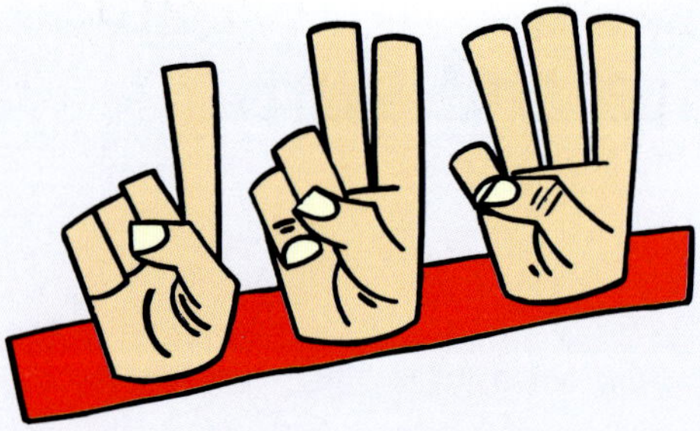

The numeral 11 is between 10 and 12.

10 ___ 12

Trace each pair of numerals, then write the numeral that comes between them.

8	___	10	9	___	11
12	___	14	7	___	9
5	___	7	8	___	10
2	___	4	1	___	3
8	___	10	12	___	14
9	___	11	11	___	13
14	___	16	13	___	15
17	___	19	16	___	18
16	___	18	18	___	20

Use the number line to show the numerals in each problem, then write the sum.

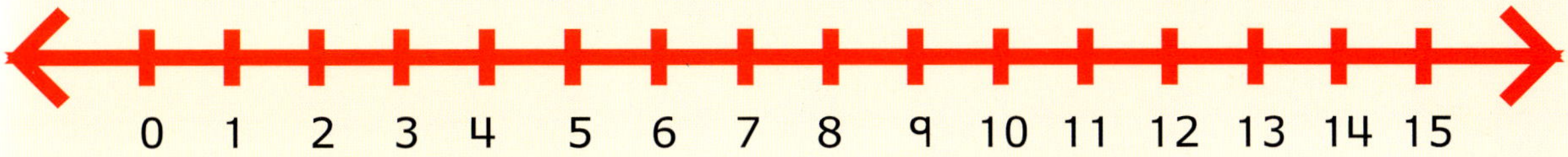

4 + 7 = ____

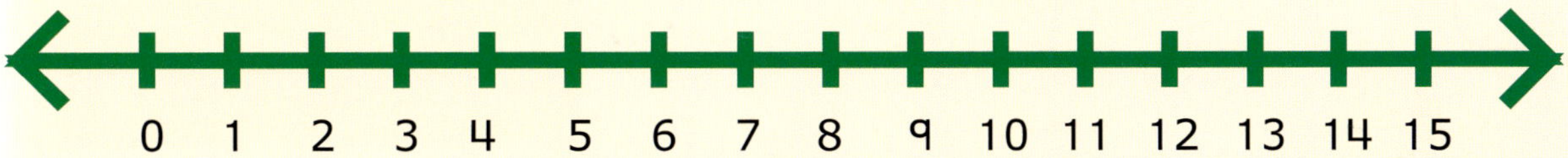

9 + 5 = ____

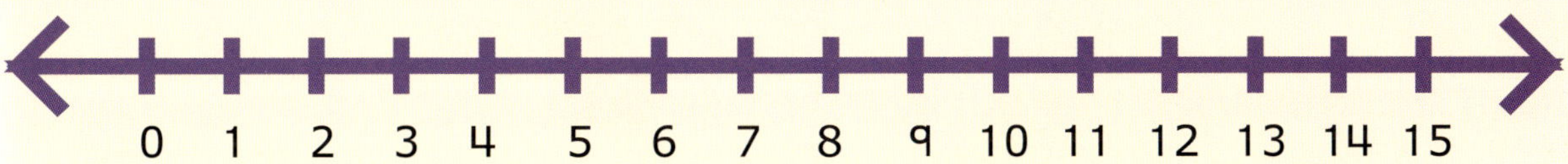

2 + 8 = ____

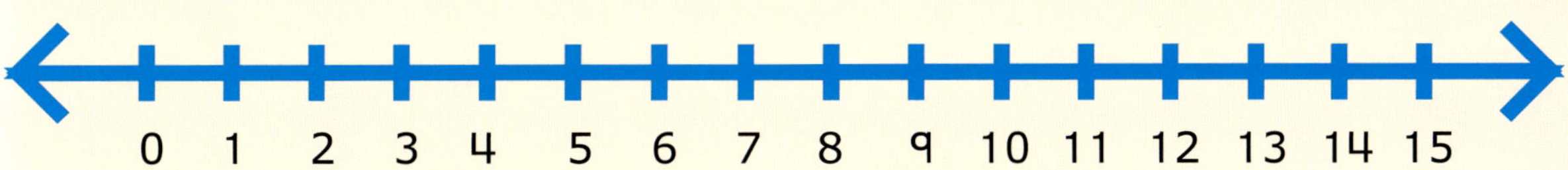

6 + 6 = ____

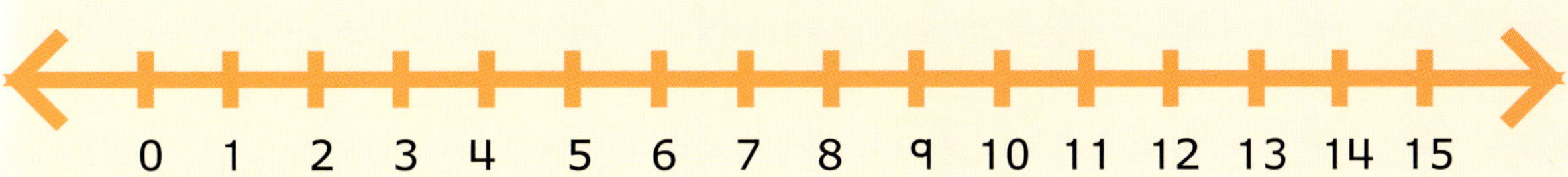

6 + 9 = ____

Draw an X on all of the shapes with less than 3 angles.

Circle the sets with exactly four corners.

Example

Do the problems. Go in order from left to right. Then, on the next page, connect the answer dots to find the mystery animal.

3 + 5	4 + 5	6 + 2	3 + 2
8 + 1	5 + 4	1 + 1	5 + 2
7 + 1	4 + 4	2 + 2	3 + 3
7 + 0	0 + 5	6 + 0	8 + 0
7 + 2	5 + 3	2 + 7	3 + 5

I live in the sea. I do not eat hay, but I love to eat shrimp.

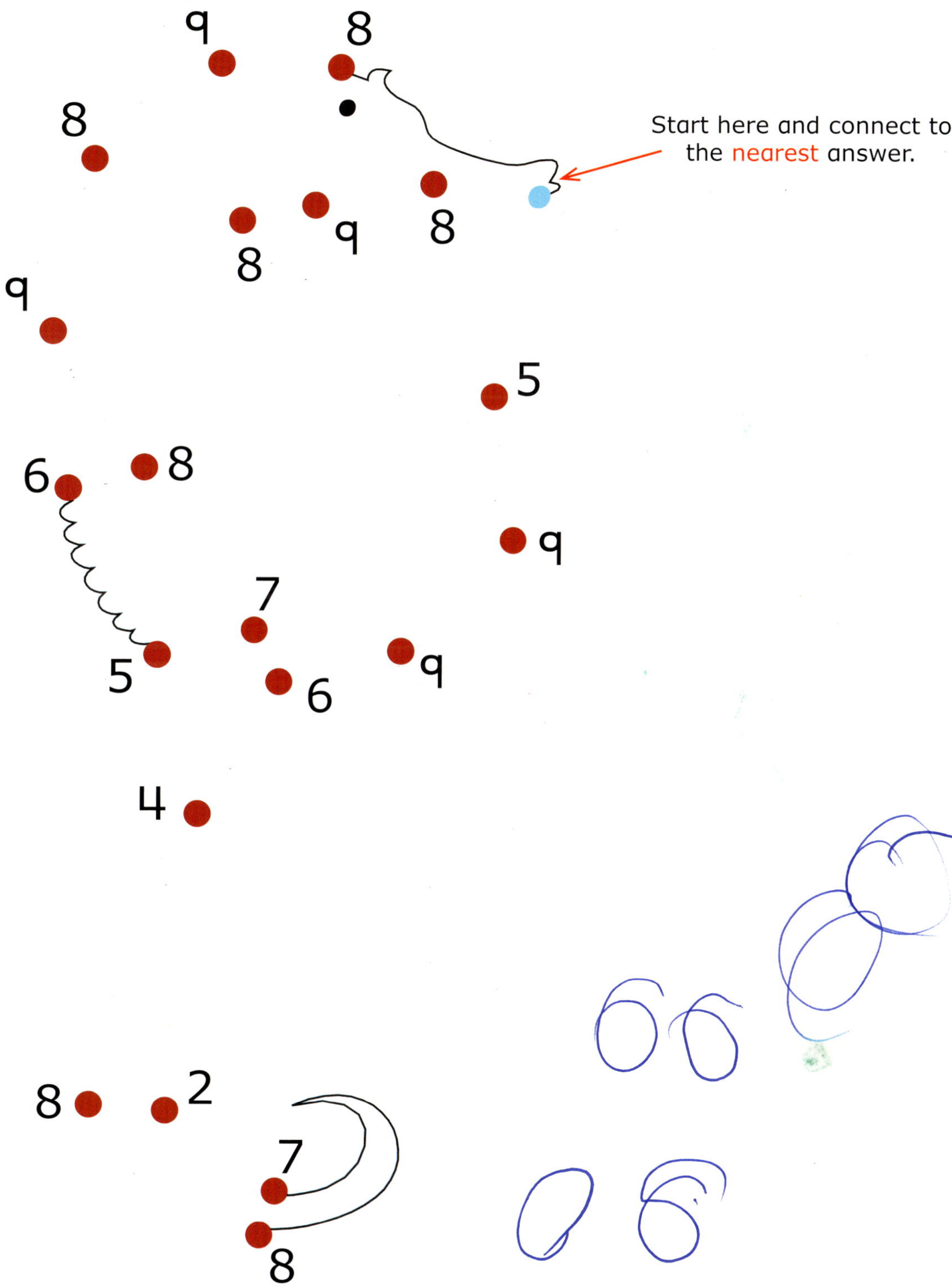

Color the mystery animal and add something new to the picture.

Circle all the sets with a square and a triangle.

Example

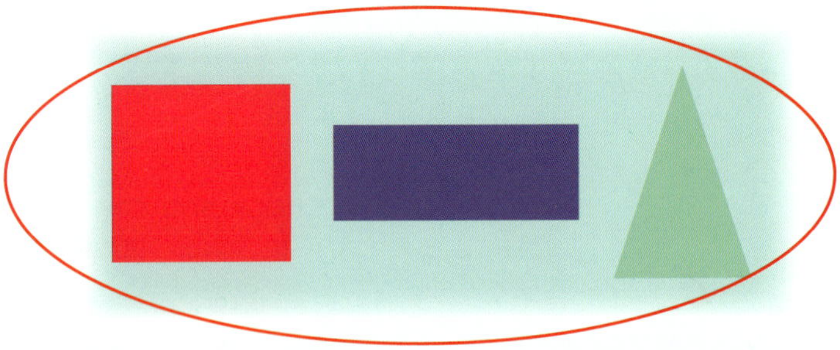

Thinker Doodles*

Chimpanzee

I play in the jungle and like to swing on tree limbs.

Connect the dots to complete the picture, then finish coloring it. Can you add something else to the picture?

*For more Thinker Doodle activities, please see our *Thinker Doodles* series.

Draw a triangle on all of the shapes without corners.

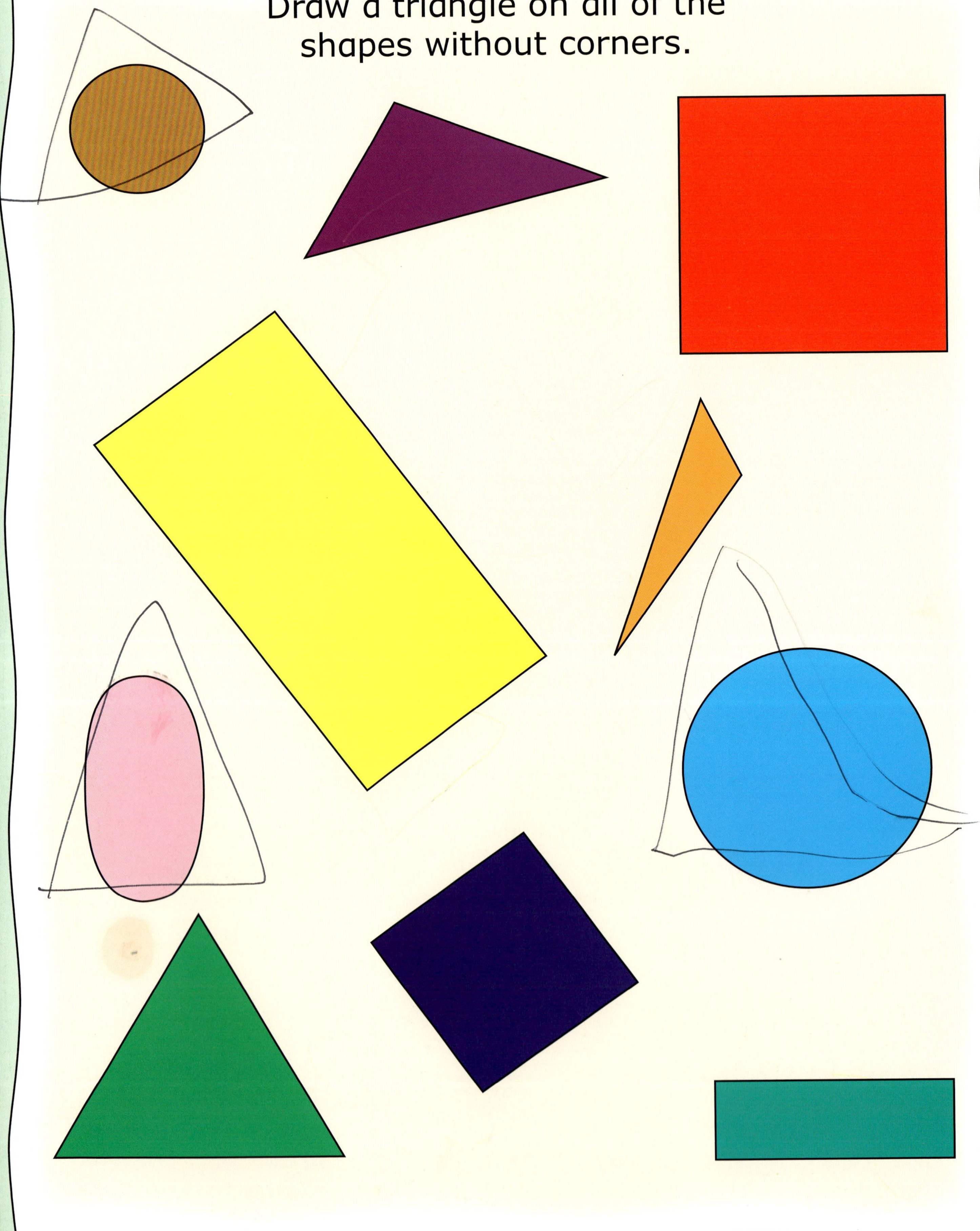

Count the fruit on the trees, then write the sum.

Draw line segments to match each name with its shape.

bicycle tire

refrigerator

pizza box

bathroom sink

football

mountain

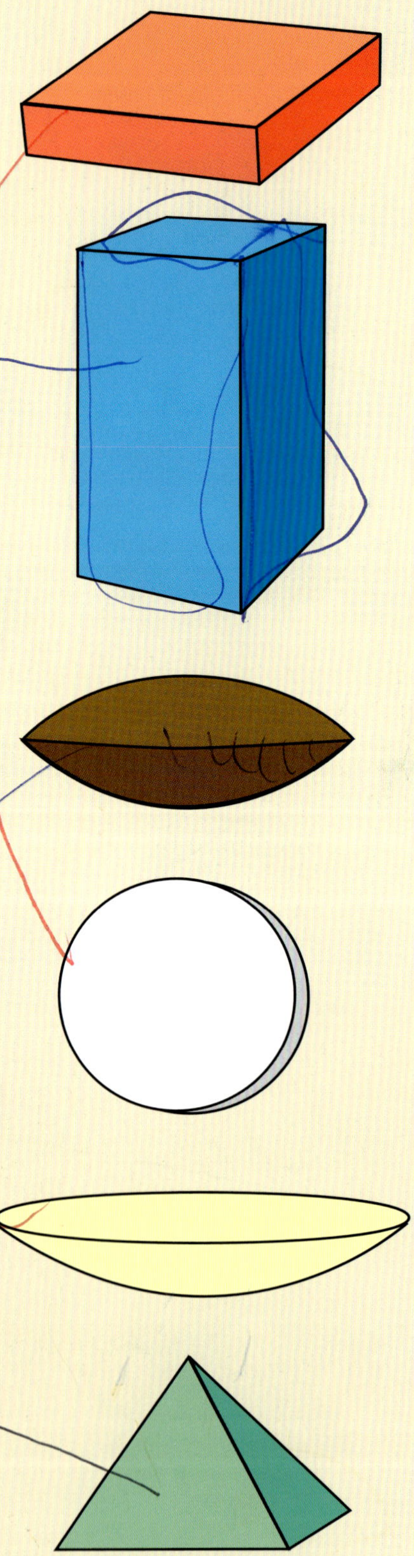

Note: Accept any answer that can be rationally justified.

Put an X over the thing in each set that does not belong, then explain your answer.

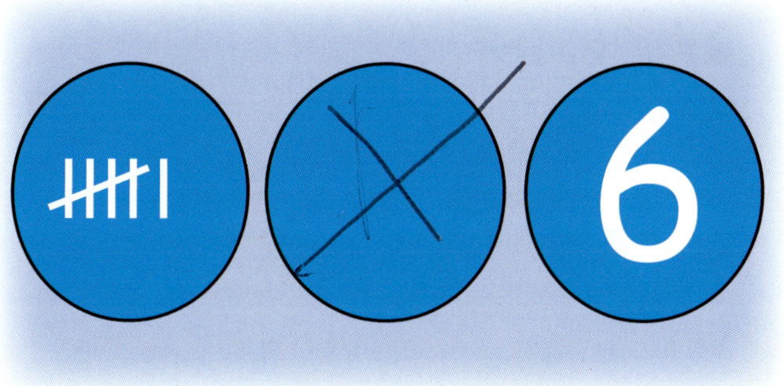

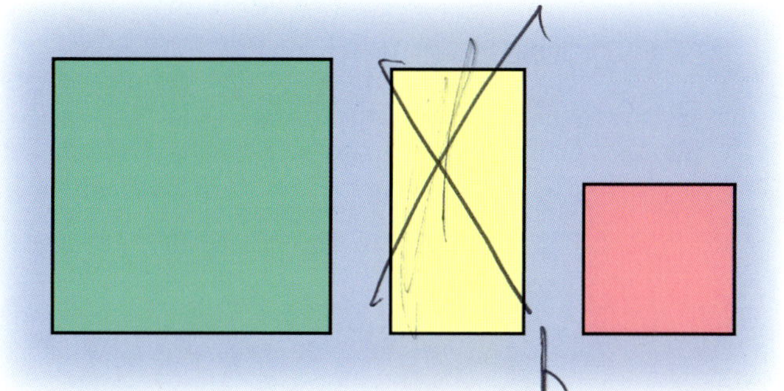

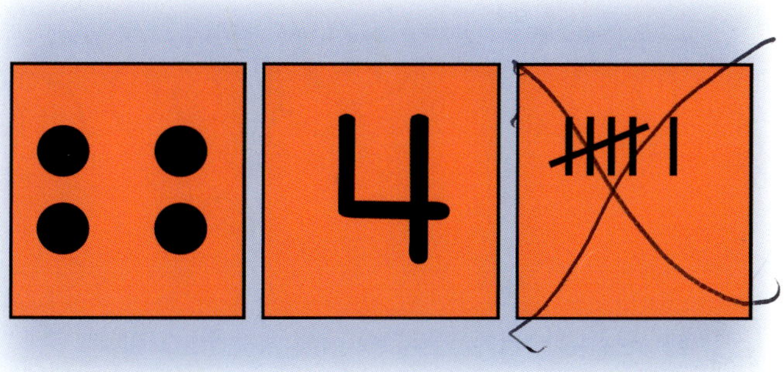

If six more balls are tossed onto the ball field, how many balls would there be on the field? ________

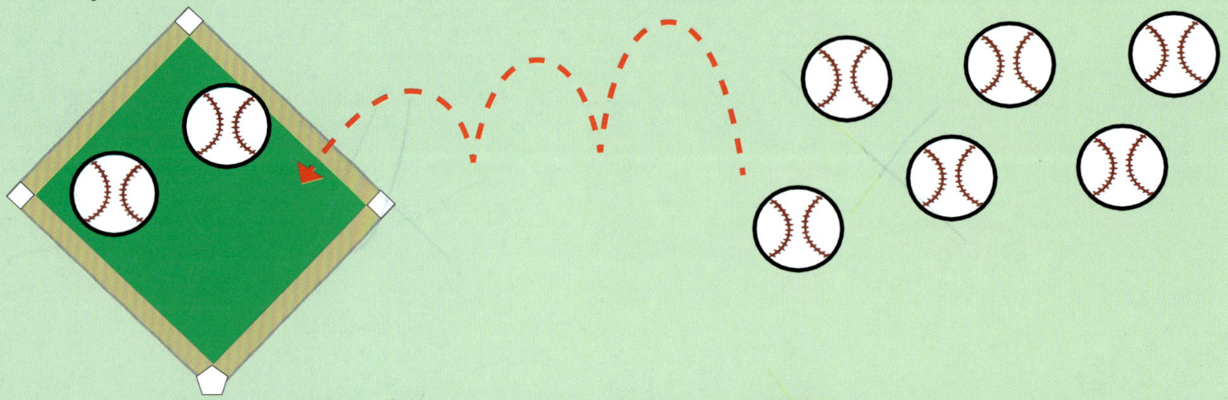

________ + ________ = ________

If ten more balls are tossed onto the ball field, how many balls would there be on the field? ________

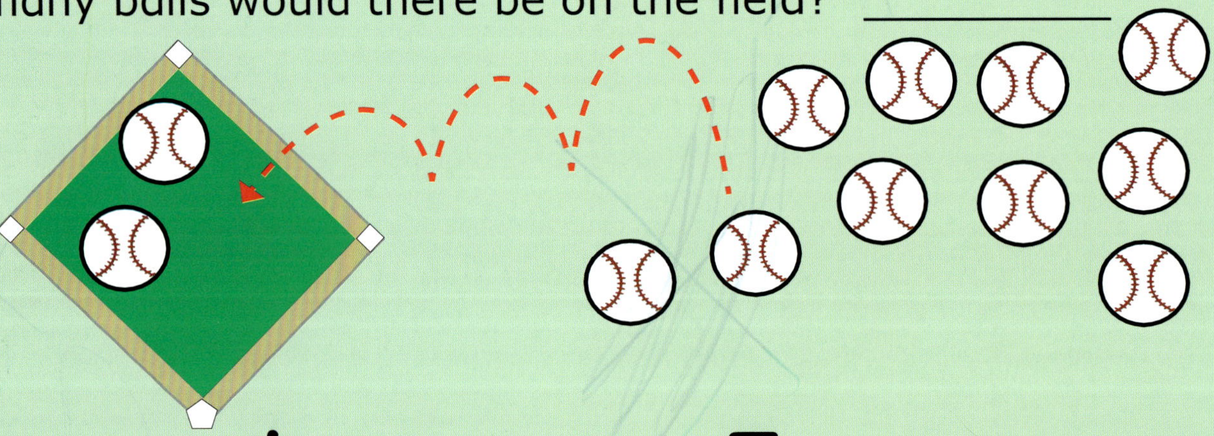

________ + ________ = ________

If twelve more balls are tossed onto the ball field, how many balls would there be on the field? ________

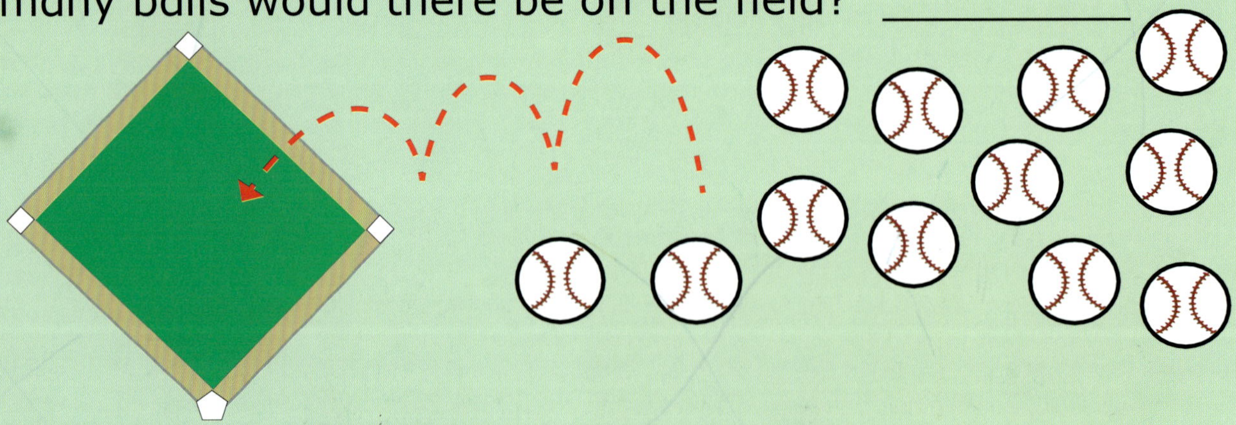

________ + ________ = ________

THINKER DOODLES*

1. Look at each dinosaur above, then find its unfinished picture below. Use a pencil to draw in all the missing parts.

2. Use 3 colors to color all the dinosaurs with 5 upper teeth.
3. Use 2 colors to color all the dinosaurs with a small black dot at the end of their tail.
4. Use 4 colors to color all the remaining dinosaurs.

*For more Thinker Doodles activities, please see our *Thinker Doodles* series.

Complete the number sentences.

 +  = 

________ + ________ = ________

 +  = 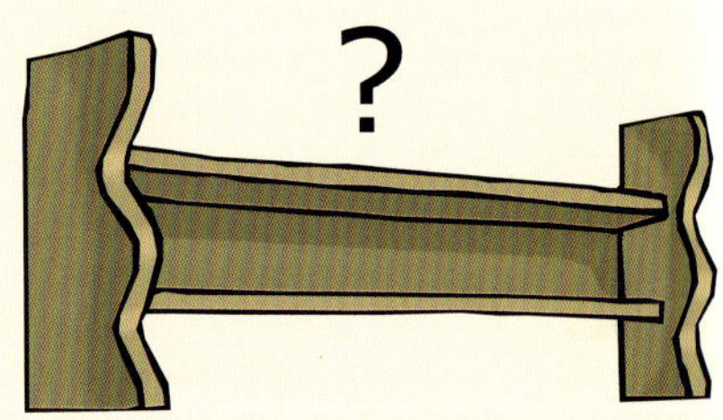

________ + ________ = ________

 +  = 

________ + ________ = ________

Write the number of cents in each set.

 = ________¢

 = ________¢

 = ________¢

 = ________¢

 = ________¢

 = ________¢

Draw line segments to connect each item with its shape.

water pipe

green pea

washing machine

box for a baseball

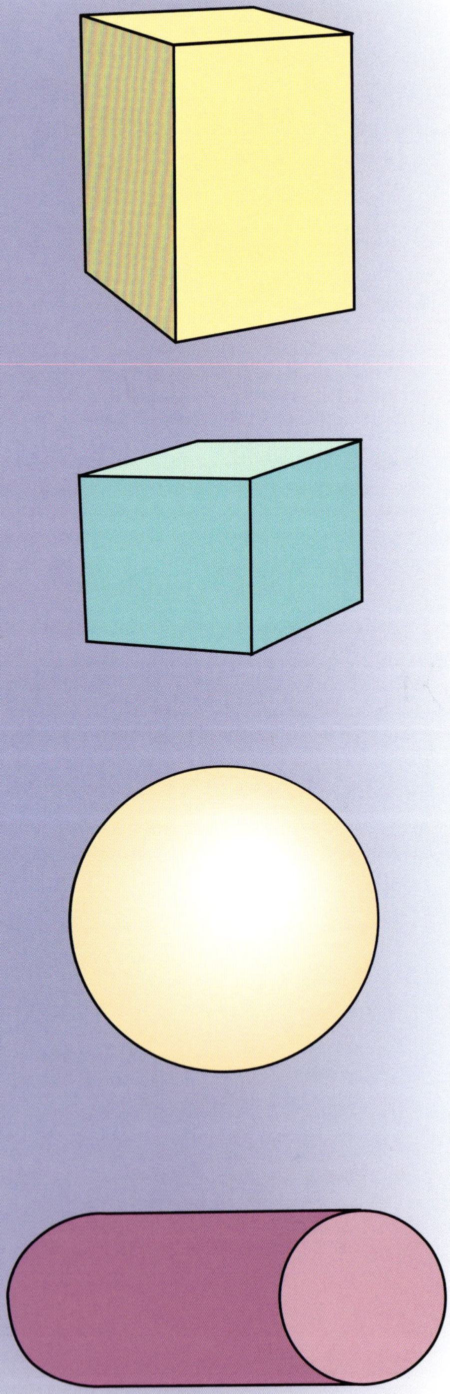

Note: Accept any answer that can be rationally justified.

Complete the number sentence for each problem.

You have your bike. Eight friends bring their bikes to your house too. How many bikes are at your house?

___ + ___ = ___

You have 2 pennies. You find 6 pennies. Later you find another penny. How many pennies do you have?

___ + ___ + ___ = ___

There are 6 books on one shelf and 2 more books on another shelf. How many books are there in all?

___ + ___ = ___

Draw, write, and say the answer for each of the following subtraction problems.

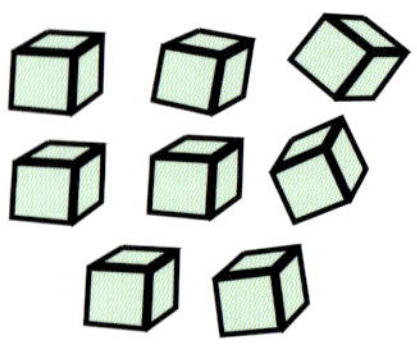

take away 3
(cross them out)

________ - ____3____ = ________

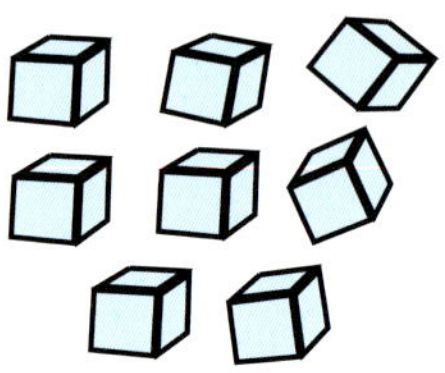

________ - ____6____ = ________

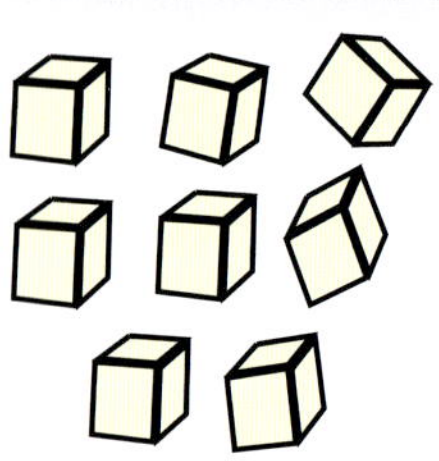

________ - ____2____ = ________

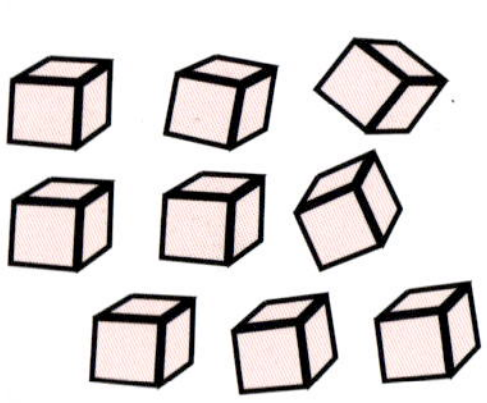

________ - ____5____ = ________

Note: It sometimes helps students if they cross out the items taken away from the original amount.

Draw, write, and say the answer for each of the following subtraction problems.

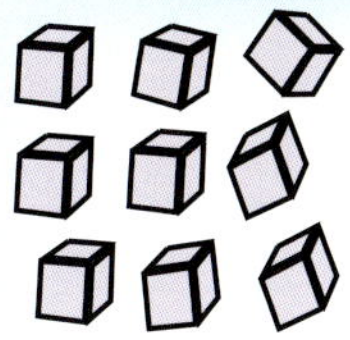

________ - ____ 9 ____ = ________

________ - ____ 0 ____ = ________

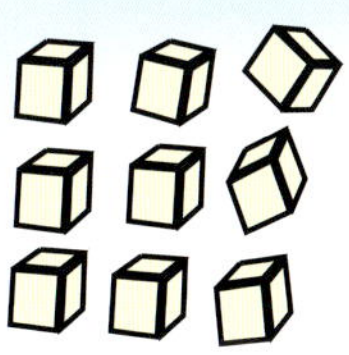

________ - ____ 4 ____ = ________

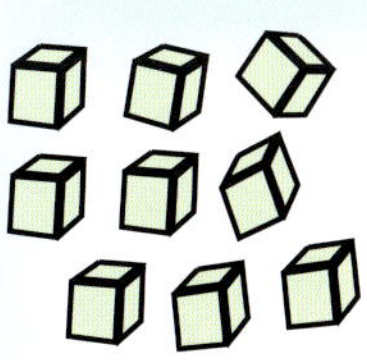

________ - ____ 5 ____ = ________

________ - ____ 3 ____ = ________

Note: It sometimes helps students if they cross out the items taken away from the original amount.

Something that is split into 2 equal parts has 2 halves. → $\frac{1}{2}$ | $\frac{1}{2}$

Something that is split into 3 equal parts has 3 thirds. → $\frac{1}{3}$ | $\frac{1}{3}$ | $\frac{1}{3}$

Something that is split into 4 equal parts has 4 fourths. → $\frac{1}{4}$ | $\frac{1}{4}$ | $\frac{1}{4}$ | $\frac{1}{4}$

Complete the fractions.

$\frac{1}{2}$ | $\frac{1}{\quad}$

$\frac{1}{\quad}$ | $\frac{1}{\quad}$ | $\frac{1}{\quad}$

$\frac{1}{\quad}$ | $\frac{1}{\quad}$ | $\frac{1}{\quad}$ | $\frac{1}{\quad}$

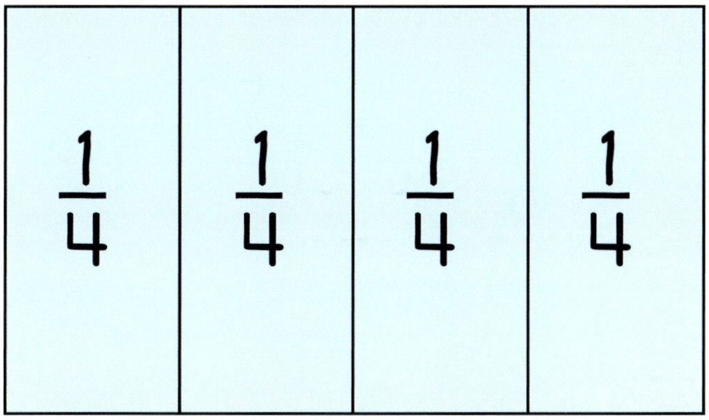

Something split into four <u>equal</u> <u>parts</u> is split into <u>fourths</u>.

Put an X on the objects split into fourths.

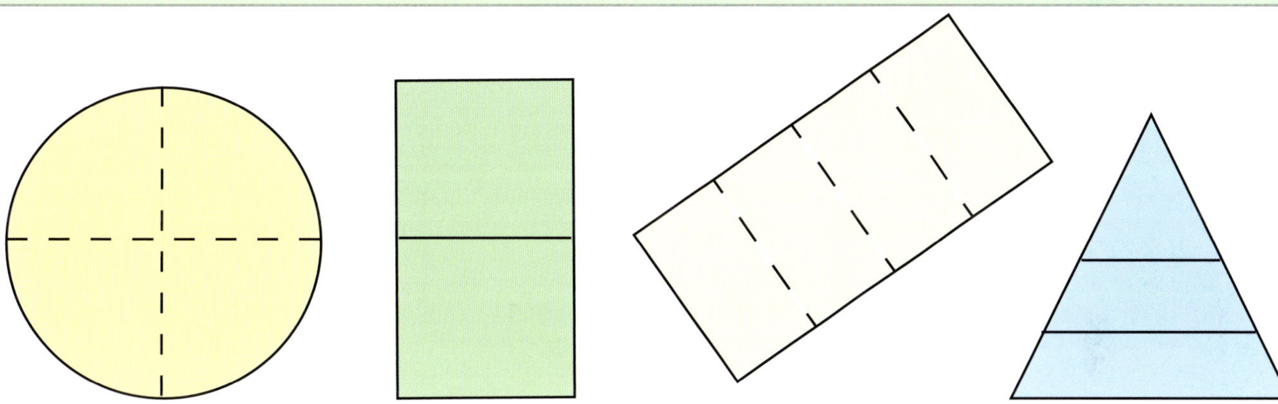

Color $\frac{1}{4}$ of each object.

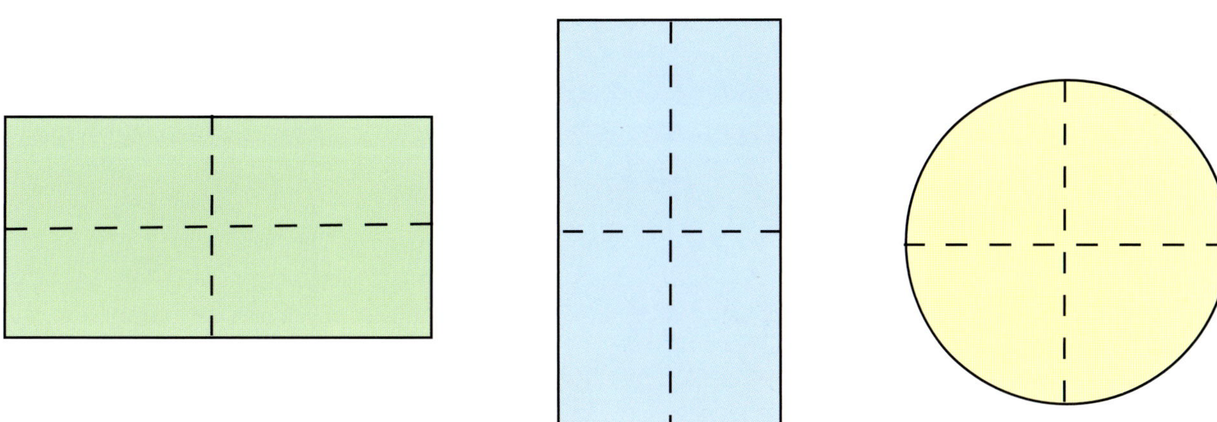

Complete the missing fractions.

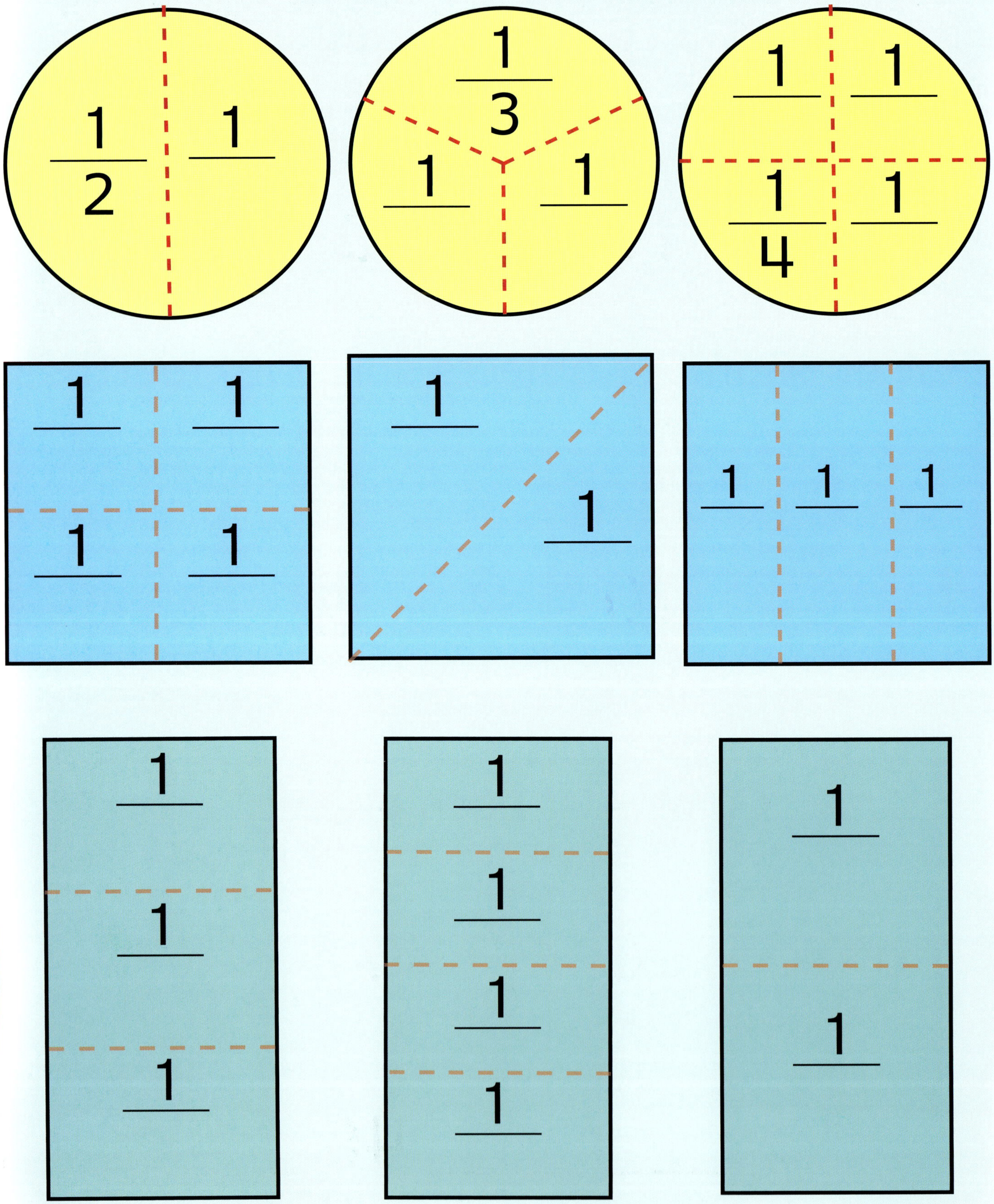

Use the number line to identify each missing numeral in the sequence.

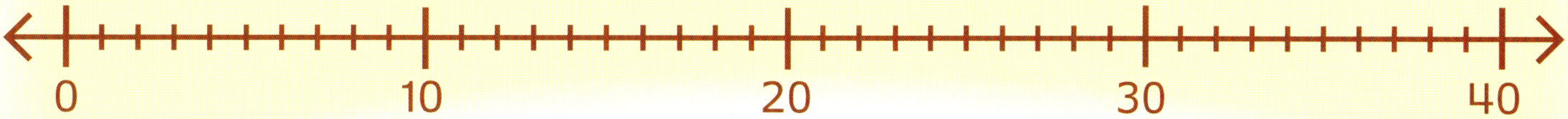

10 ____ 30

20 ____ 40

20 30 ____

10 ____ 30 ____ 50

10 20 ____ 40 ____

10 ____ ____ 40 ____

Circle each group that equals 25¢.

A quarter  is worth 25 cents (25¢).

Example

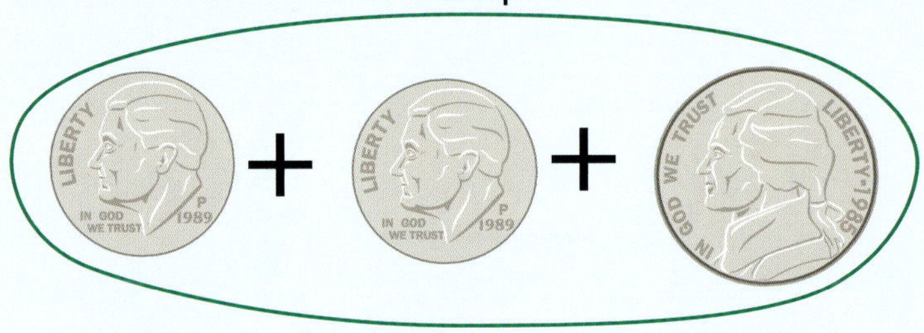

10¢ + 10¢ + 5¢ = 25¢

Trace each numeral, then continue the pattern.

1 4 5 1 4 5 1 4 5 ____ ____ ____

4 1 5 4 1 5 4 1 5 ____ ____ ____

2 3 5 2 3 5 2 3 5 ____ ____ ____

3 2 5 3 2 5 3 2 5 ____ ____ ____

0 5 5 0 5 5 0 5 5 ____ ____ ____

5 0 5 5 0 5 5 0 5 ____ ____ ____

Fill in the missing numerals to complete the pattern.

5	10		20			
	20	30	40			
	10	15	20			
10		30		50		
5		15			30	
40	50					100

Touch each dime and count by tens to complete each number sentence.

 +  +  + 

say: 10 → 20 → 30 → 40

Example

10 + 10 + 10 + 10 = 40

Touch each coin and count.

 +  + 

say: 10 → 20 → 25 → = ?______

Touch each coin and count.

How much?______

Trace the numerals, then write the next three numerals that would continue the pattern.

4 5 9 4 5 9 4 5 9 ____ ____ ____

5 4 9 5 4 9 5 4 9 ____ ____ ____

2 5 7 2 5 7 2 5 7 ____ ____ ____

3 5 8 3 5 8 3 5 8 ____ ____ ____

8 8 0 8 8 0 8 8 0 ____ ____ ____

8 0 8 8 0 8 8 0 8 ____ ____ ____

A puppet theatre has 8 red puppets and 9 blue puppets. Create a bar graph that shows this.

A painter has 9 brushes and 8 pails. Create a bar graph to show this.

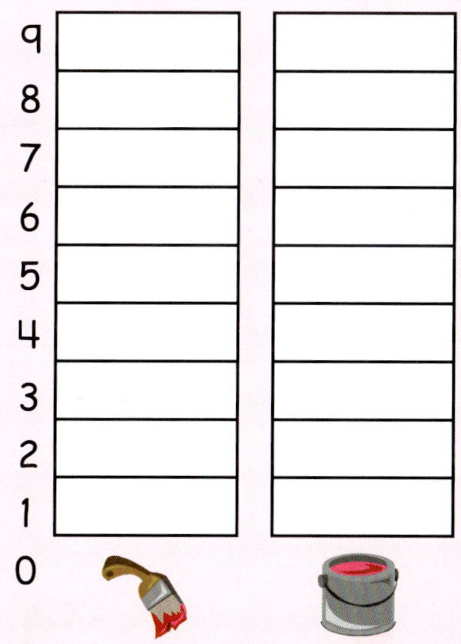

There are 9 boxes of popcorn and 9 sodas left at the movies. Create a bar graph that shows this.

Do the problems. Go in order from left to right. Then, on the next page, connect the answer dots to find the mystery animal.

8 +1	5 +4	3 +2	5 +2
3 +5	4 +5	6 +2	7 +1
1 +6	3 +5	6 +3	3 +6
7 +2	2 +7	5 +3	3 +5
3 +3	4 +4	2 +2	1 +1
7 +0	0 +5	6 +0	8 +0

Draw the tally marks to find the solution to each sentence. Then write the number sentence.

||||||||| − | = ____

____ − ____ = ____

|||||||| − || = ____

____ − ____ = ____

|||||||| − ||||| = ____

____ − ____ = ____

||||/ |||| − ||||| = ____

____ − ____ = ____

||||/ |||| − |||| = ____

____ − ____ = ____

Circle the coins you would use to make the amount in the green circle.

12¢ =

15¢ =

9¢ =

11¢ =

Use the number line to show the solution to each number sentence.

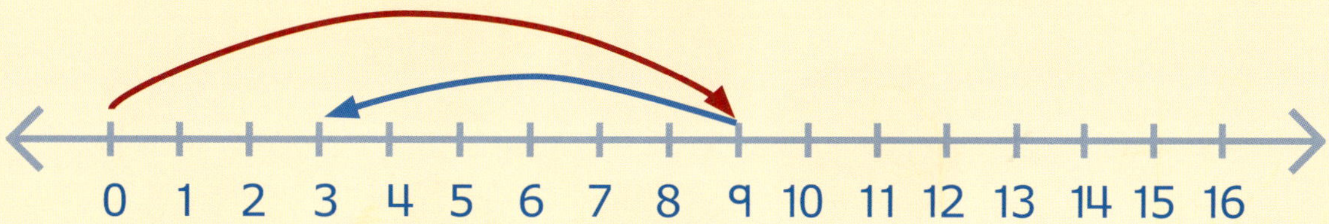

9 - 6 = 3

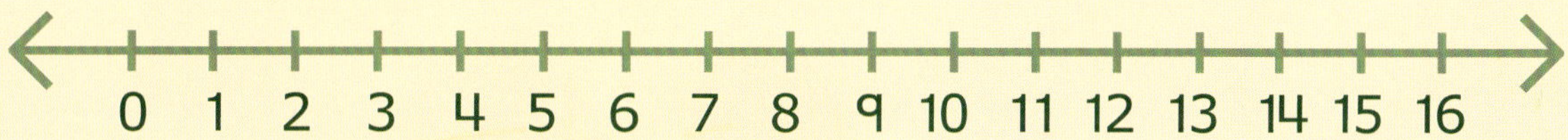

9 - 6 = ____

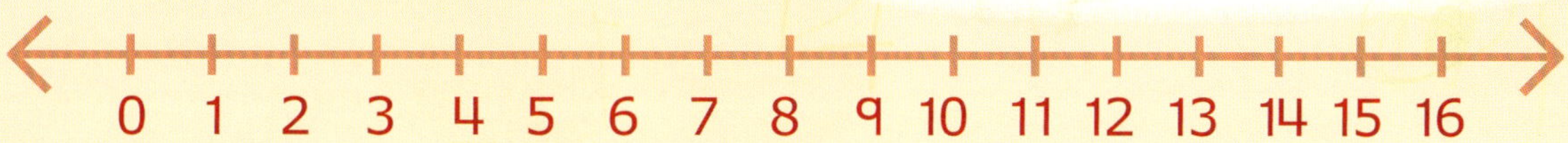

8 - 6 = ____

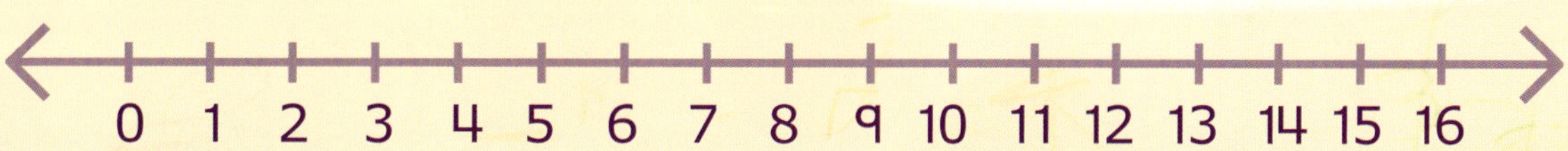

8 - 2 = ____

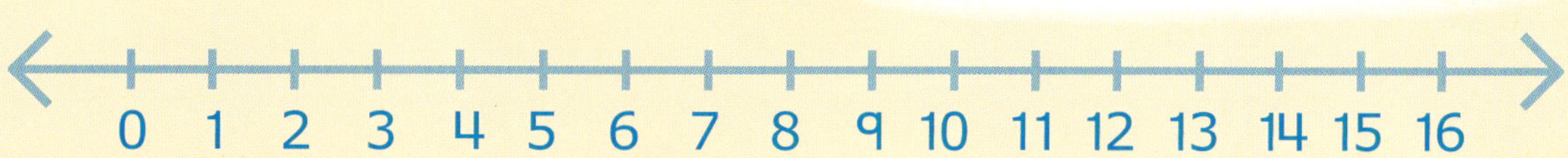

9 - 4 = ____

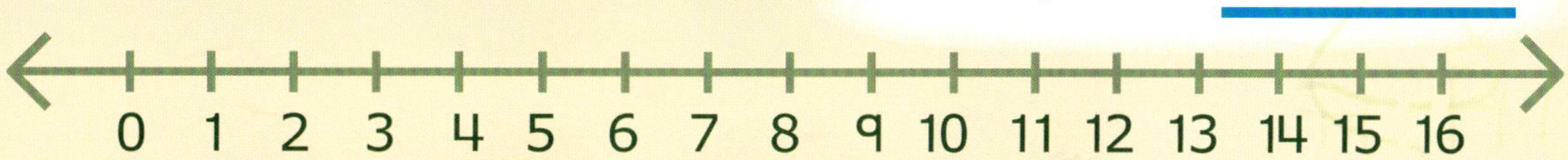

9 - 7 = ____

Circle the coins you would use to make the amount in the blue circle.

14¢ =

12¢ =

8¢ =

13¢ =

Use the number line to show the solution to each number sentence.

0 1 2 3 4 5 6 7 8 9 10 11 12 13 14 15 16

9 - 9 = ______

0 1 2 3 4 5 6 7 8 9 10 11 12 13 14 15 16

9 - 2 = ______

0 1 2 3 4 5 6 7 8 9 10 11 12 13 14 15 16

7 - 3 = ______

0 1 2 3 4 5 6 7 8 9 10 11 12 13 14 15 16

8 - 6 = ______

0 1 2 3 4 5 6 7 8 9 10 11 12 13 14 15 16

7 - 4 = ______

0 1 2 3 4 5 6 7 8 9 10 11 12 13 14 15 16

8 - 7 = ______

Do the problems. Go in order from left to right. Then, on the next page, connect the answer dots to find the mystery animal.

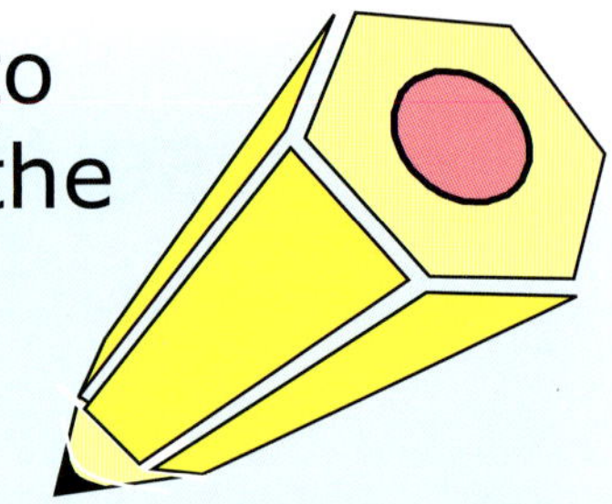

3 − 2	4 − 4	6 − 1	7 − 4
8 − 5	5 − 4	3 − 2	5 − 3
7 − 2	8 − 6	5 − 4	9 − 5
1 − 1	3 − 3	6 − 5	4 − 2
7 − 2	3 − 0	6 − 5	8 − 4

I am a good swimmer and live in the weeds next to water. I love to eat bugs.

Start & finish here. Connect to the nearest answer.

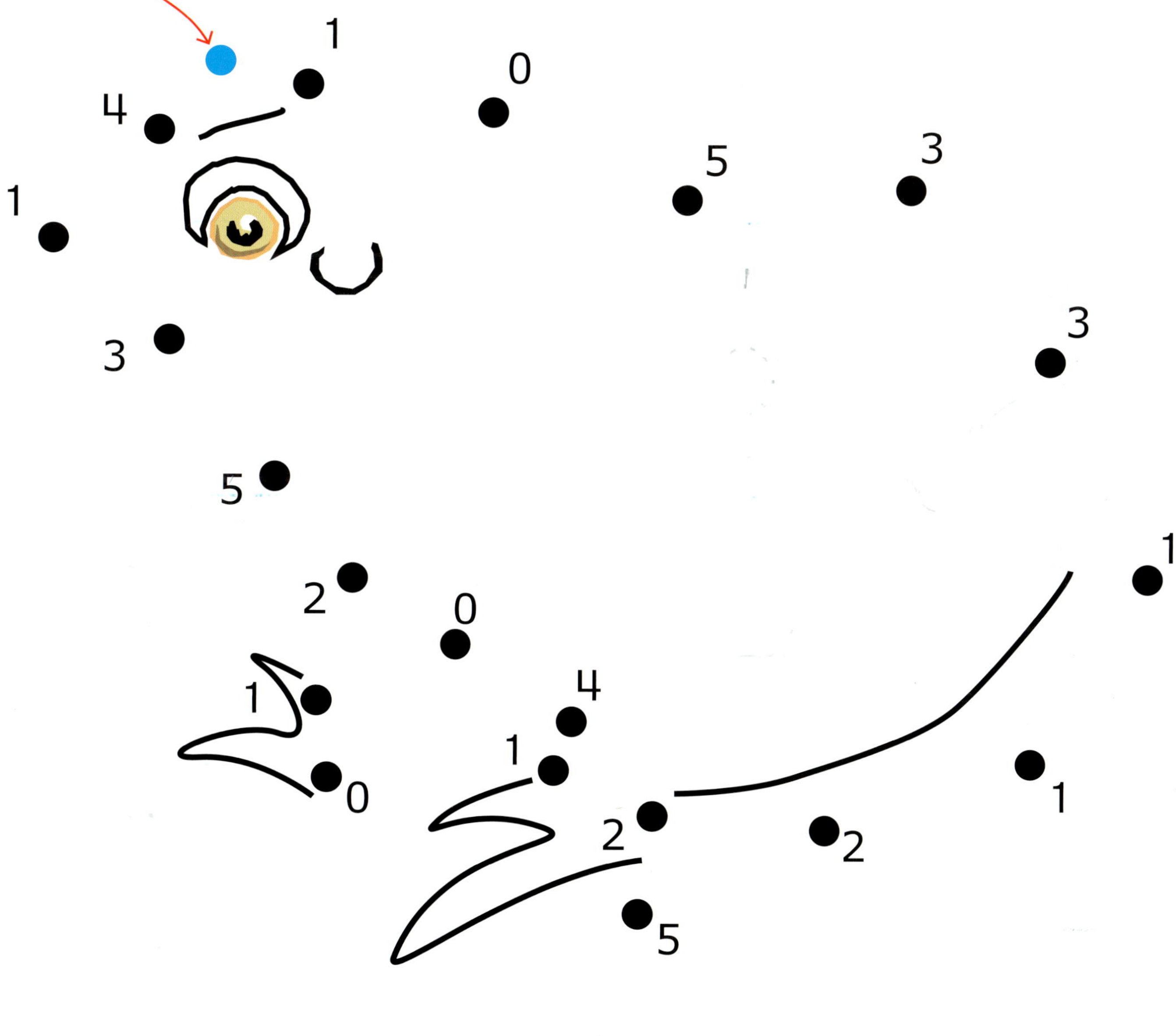

Color the mystery animal and add something new to the picture.

Thinker Doodles*

Lion

I roar loudly and am called the king of the jungle.

Connect the dots to complete the picture, then finish coloring it. Can you add something else to the picture?

*For more Thinker Doodle activities, please see our *Thinker Doodles* series.

MIND BENDERS® *

DIRECTIONS: Fill in the chart using Y for yes or N for no as you solve the puzzle.

Tim, Pam, Jill, and Todd all have change from the store. Fnd each person's change.

1. Jill has curly hair and more cents than Pam, Todd, and Tim.
2. Tim has straight hair and more cents than Pam, but less than Todd.

* For more Mind Benders® activities, please see our *Mind Benders®* series.

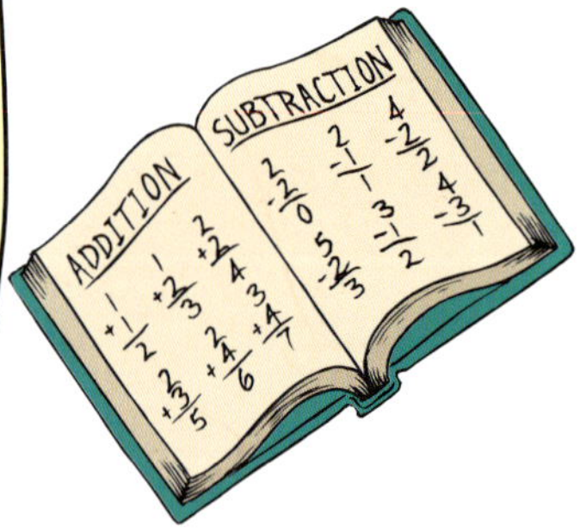

Do the problems. Go in order from left to right. Then, on the next page, connect the answer dots to find the mystery object.

9 - 2 = ____ 8 - 2 = ____

7 - 2 = ____ 9 - 8 = ____

8 - 7 = ____ 9 - 6 = ____

9 - 3 = ____ 7 - 6 = ____

9 - 2 = ____ 8 - 5 = ____

9 - 1 = ____ 9 - 5 = ____

8 - 1 = ____ 8 - 3 = ____

8 - 5 = ____ 8 - 6 = ____

9 - 4 = ____ 8 - 2 = ____

9 - 3 = ____ 9 - 0 = ____

I have a great smell while I'm cooking. I am held with both hands when eaten, and I'm often eaten in a car.

Start & finish here. Connect to the nearest answer.

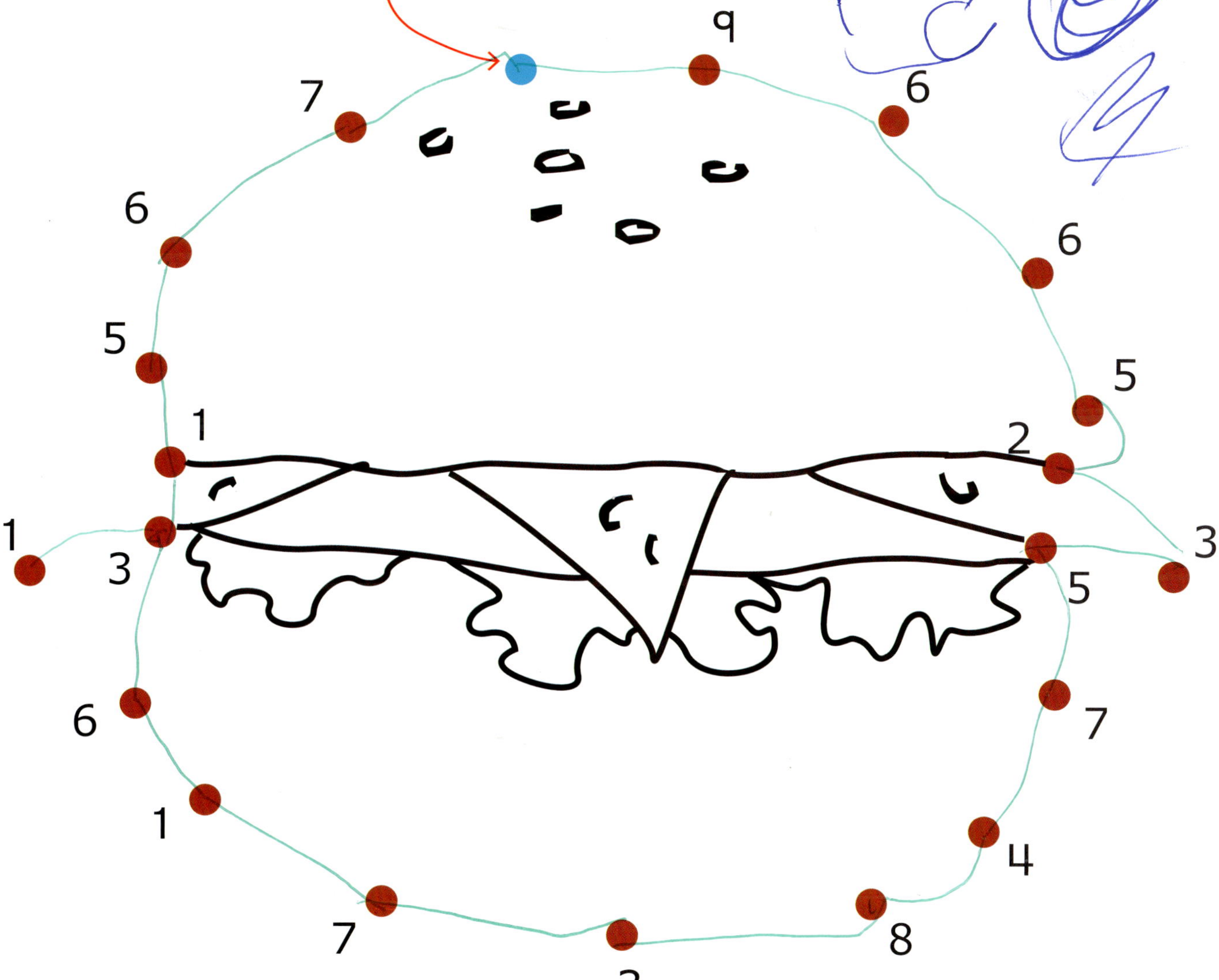

Color the mystery object and add something new to the picture.

Put an X over the thing that does not belong in each set.

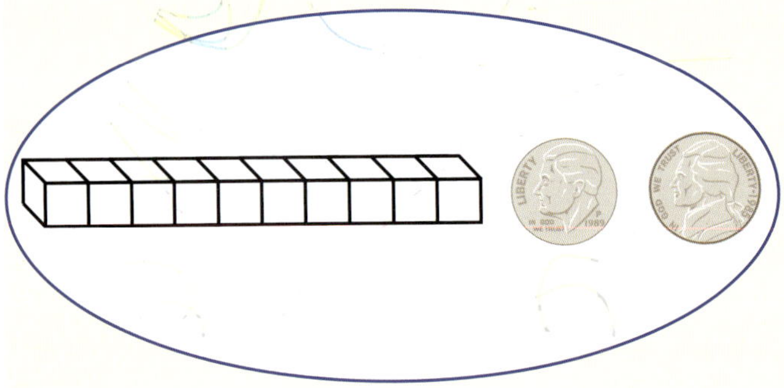

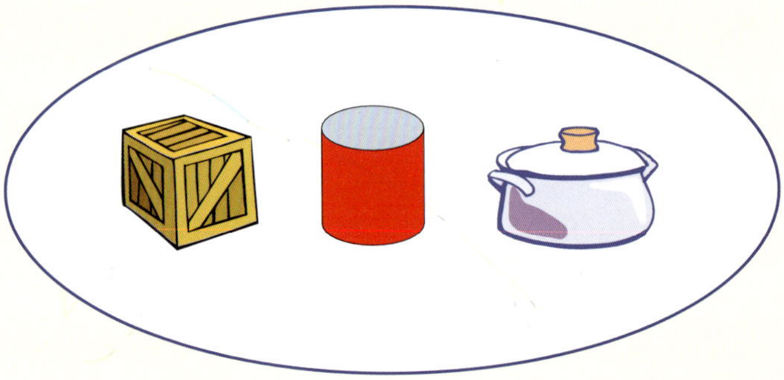

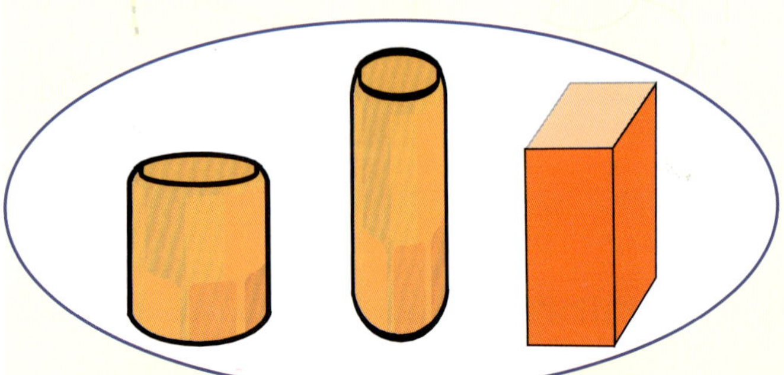

Count the books on each bookshelf, then complete the number sentence.

If you took eight books off the shelf, how many books would be left?

______ − ______ = ______

If you took three books off the shelf, how many books would be left?

______ − ______ = ______

If you took one book off the shelf, how many books would be left?

______ − ______ = ______

If you took five books off the shelf, how many books would be left?

______ − ______ = ______

Do the problems. Go in order from left to right. Then, on the next page, connect the answer dots to find the mystery object.

$8 - 2$	$5 - 3$	$3 - 1$	$5 - 4$
$7 - 4$	$8 - 2$	$6 - 1$	$3 - 2$
$5 - 2$	$8 - 4$	$6 - 2$	$8 - 3$
$7 - 1$	$2 - 1$	$6 - 4$	$4 - 1$
$3 - 2$	$7 - 4$	$4 - 2$	$7 - 5$
$5 - 3$	$6 - 1$	$1 - 0$	$6 - 3$

I am used for travel to many places, but I don't travel on the ground.

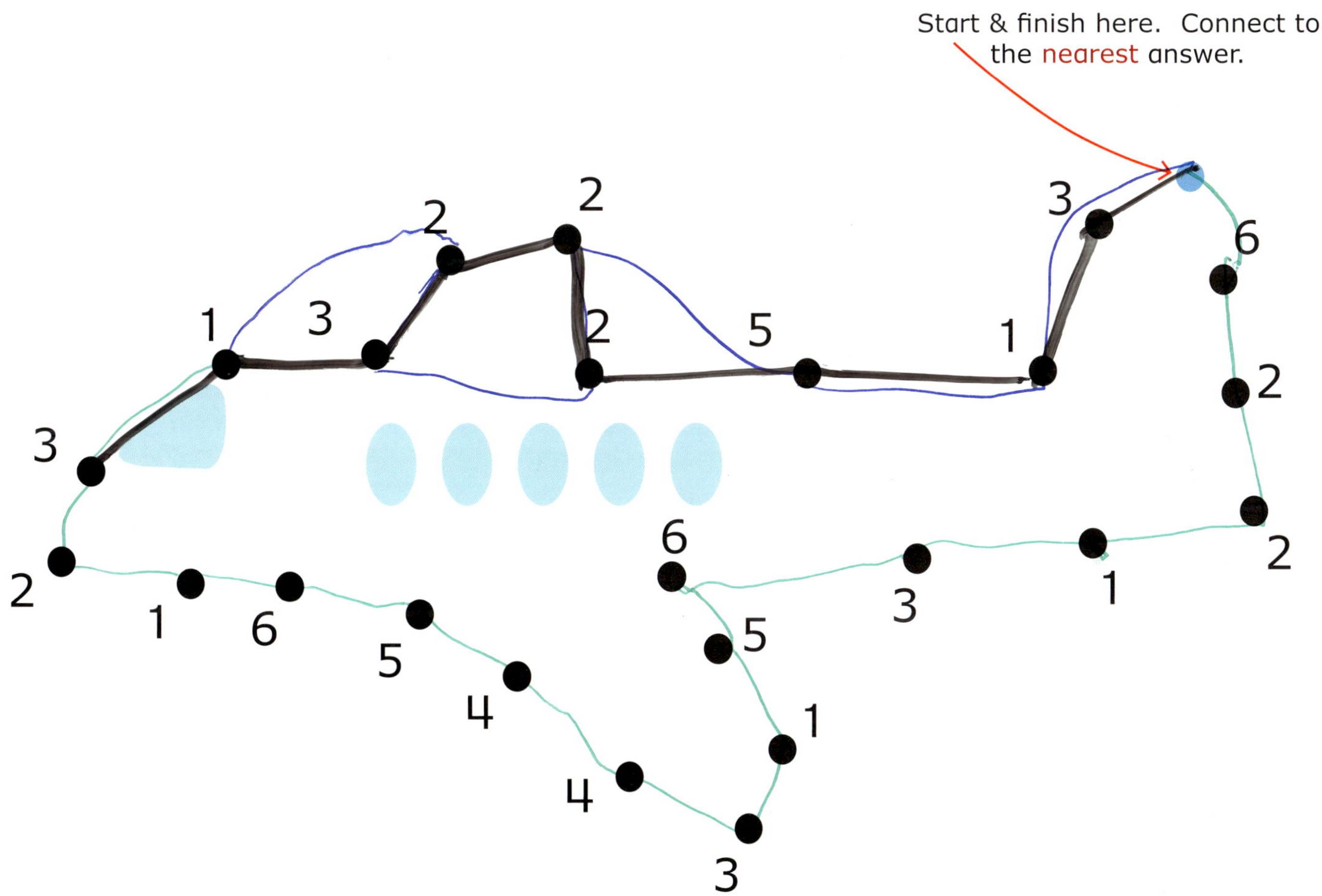

Color the mystery object and add something new to the picture.

Draw and color the objects that would continue the pattern.

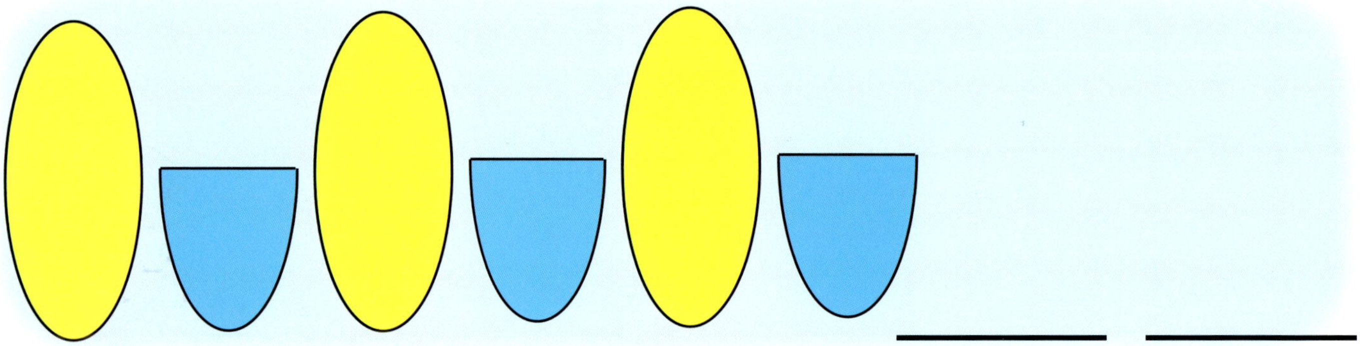

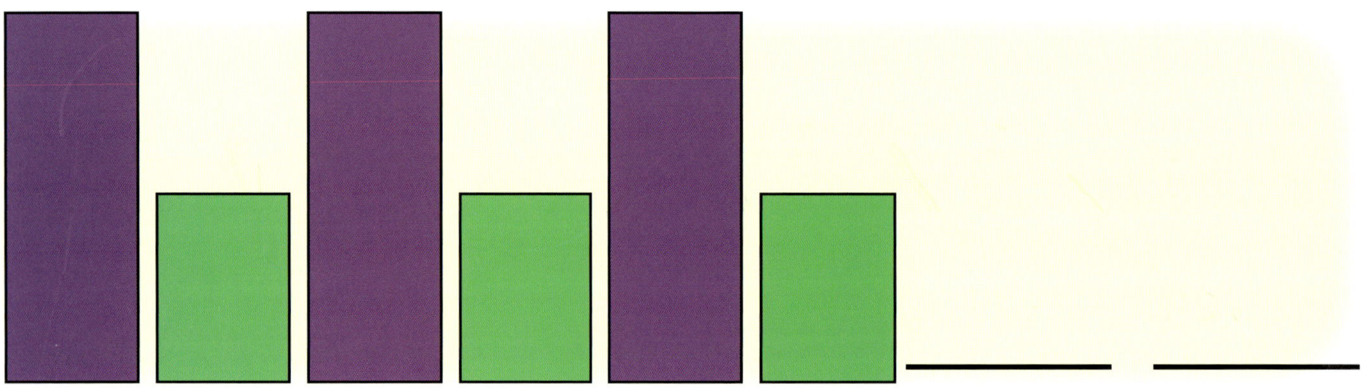

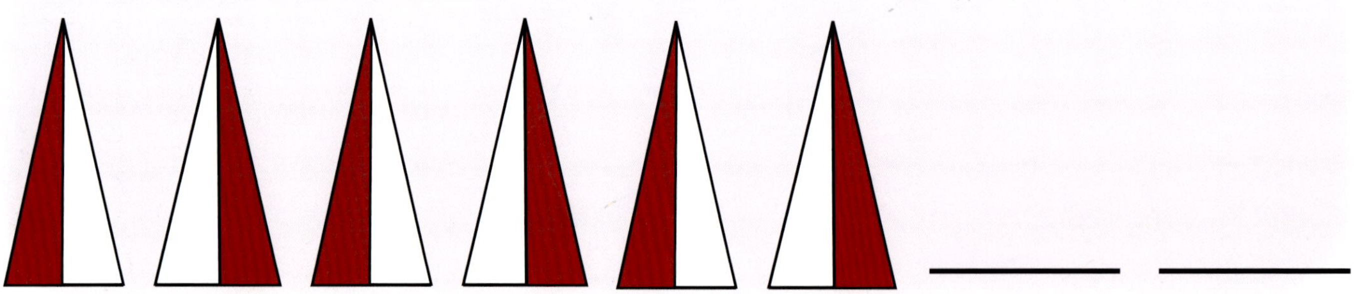

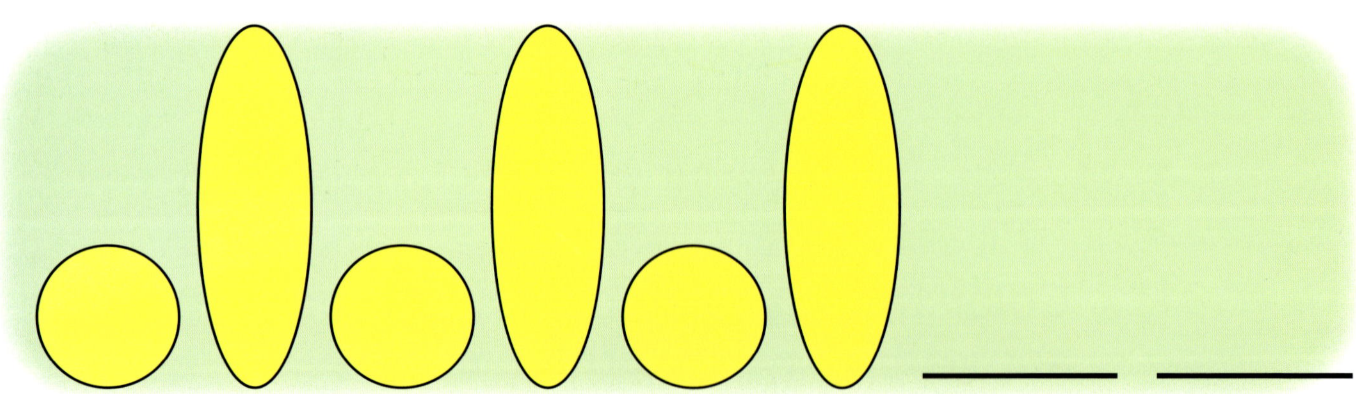

Skip and count by 2.

Two Four Six Eight Ten

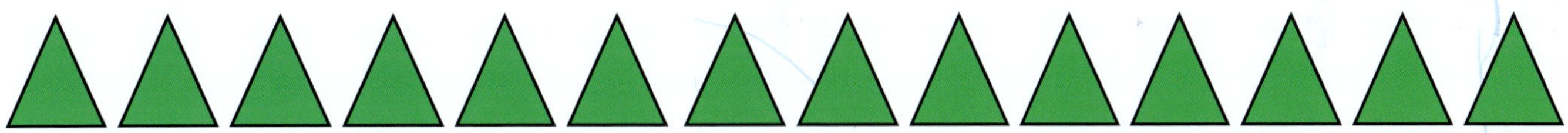

Circle sets of 2, then skip and count by 2.

How many triangles? _______

How many sets of 2? _______

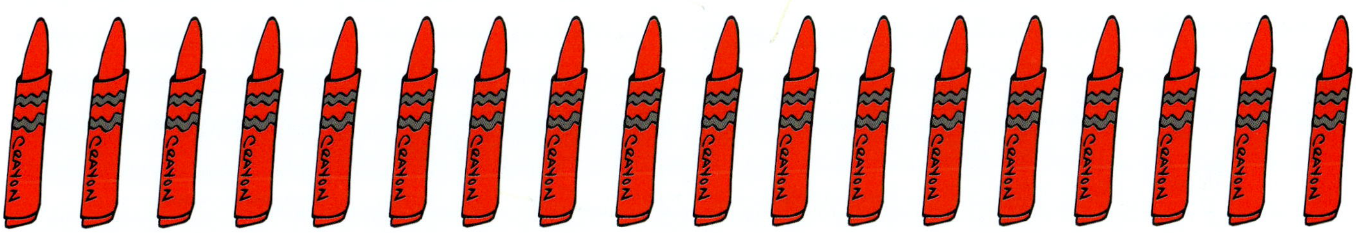

Circle sets of 2, then skip and count by 2.

How many crayons? _______

How many sets of 2? _______

Do the problems. Go in order from left to right. Then, on the next page, connect the answer dots to find the mystery animal.

6 - 5 = ______	5 - 4 = ______
7 - 6 = ______	6 - 3 = ______
8 - 7 = ______	7 - 2 = ______
8 - 5 = ______	8 - 1 = ______
9 - 0 = ______	4 - 2 = ______
5 - 1 = ______	6 - 2 = ______
4 - 4 = ______	5 - 3 = ______
7 - 5 = ______	6 - 0 = ______
8 - 0 = ______	9 - 0 = ______
5 - 4 = ______	0 - 0 = ______

I have a curly tail and live in a pen. I like to eat corn and lay in the sun.

Start & finish here. Connect to the nearest answer.

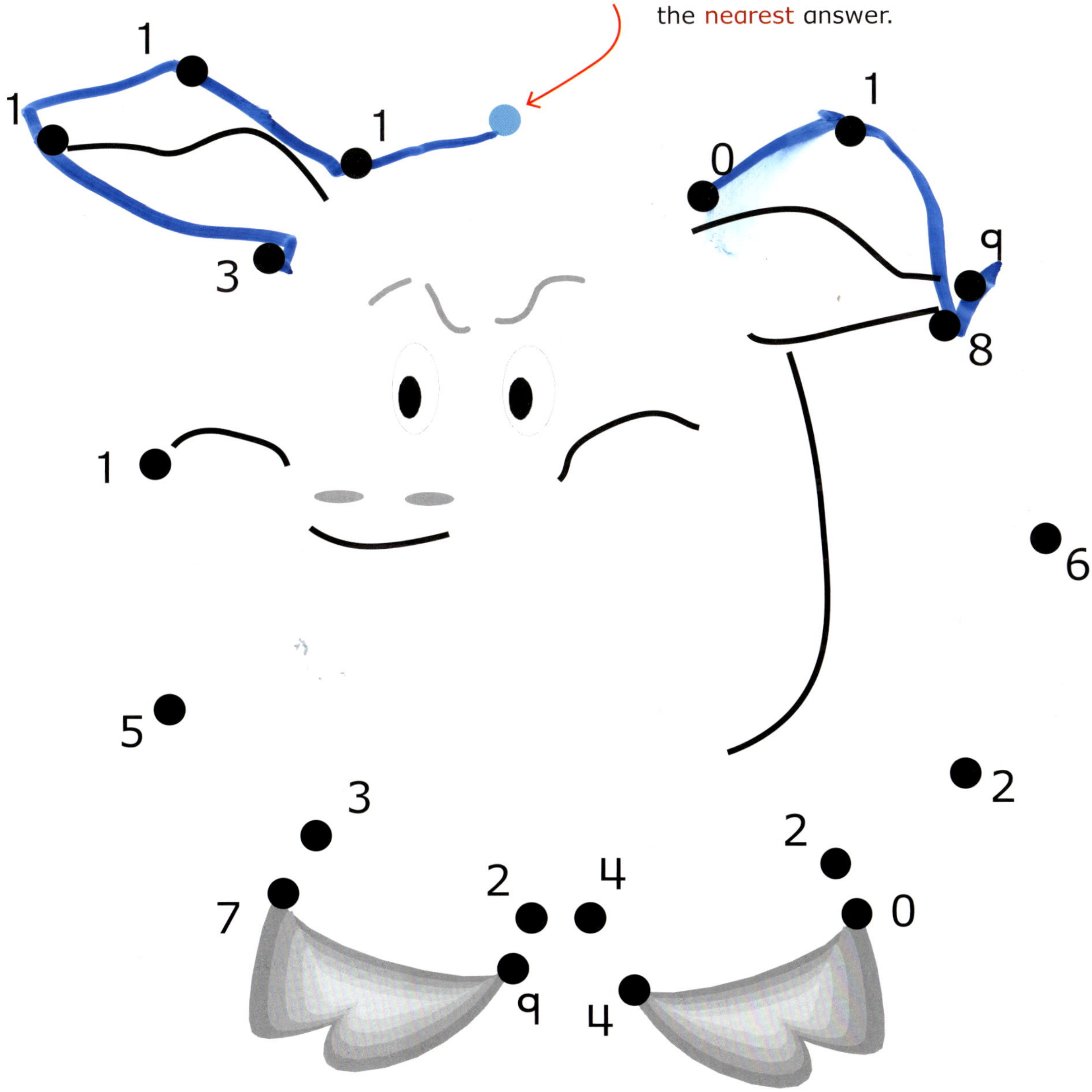

Color the mystery animal and add something new to the picture.

Draw and color the objects that would continue the pattern.

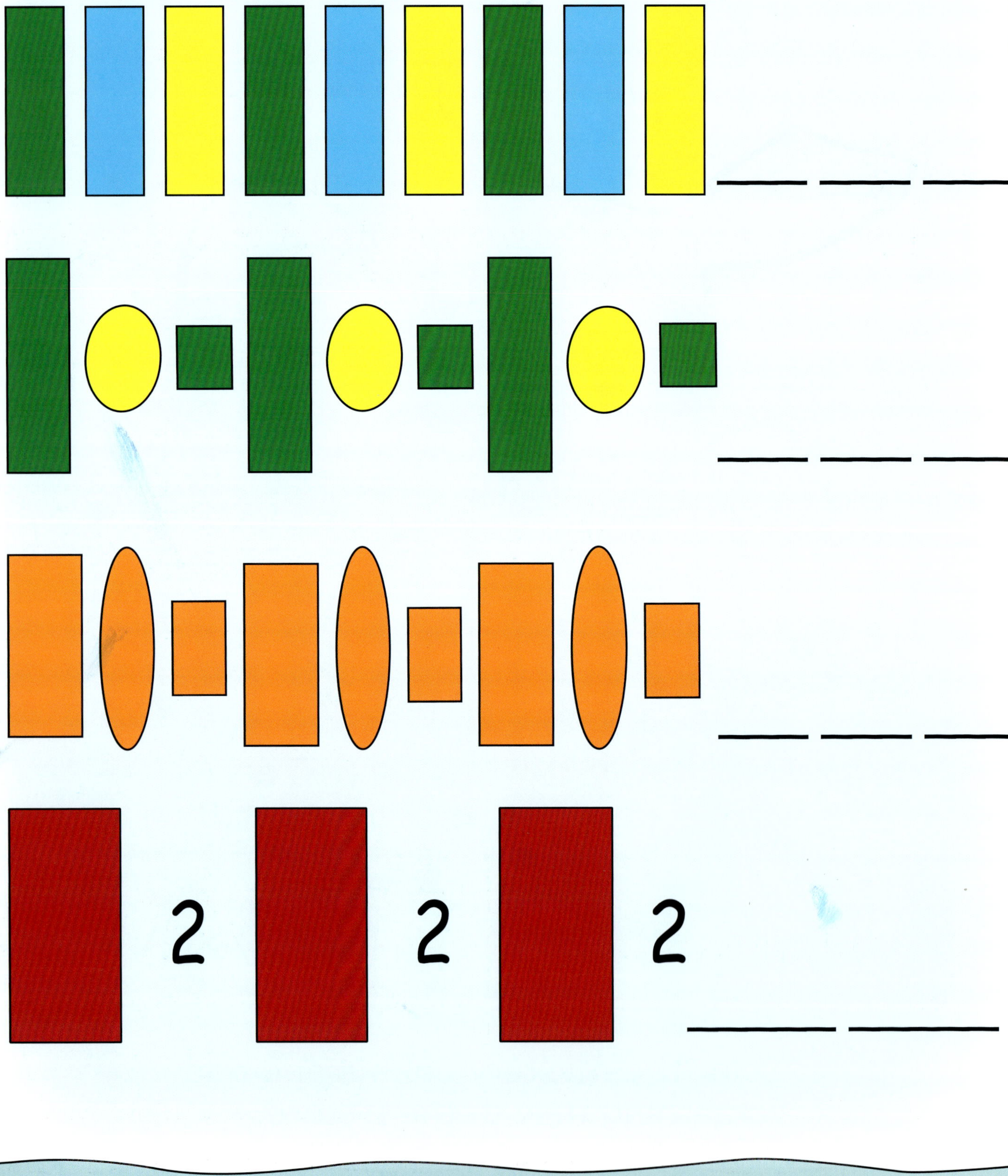

Count the toys on each shelf, then write the number sentence.

If your mom bought one toy rabbit off the shelf, how many would be left?

________ – ________ = ________

If your mom bought three toy bears off the shelf, how many would be left?

________ – ________ = ________

If your mom bought two toy clowns off the shelf, how many would be left?

________ – ________ = ________

If your mom bought one toy truck off the shelf, how many would be left?

________ – ________ = ________

Tracing the path from zero to the selected value by showing the numbers in between. For example, 18 would be located by showing:

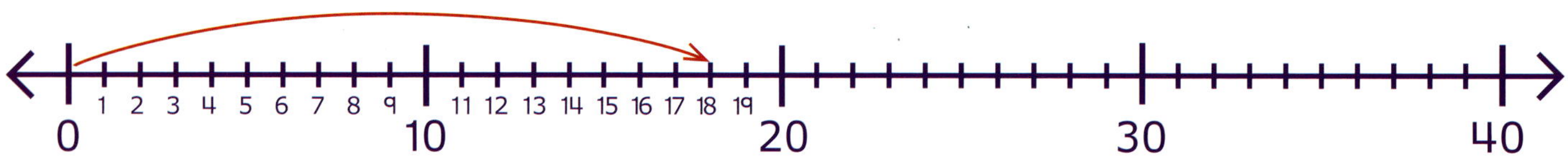

Locate 15 on the number line by tracing the path from 0 to 15.

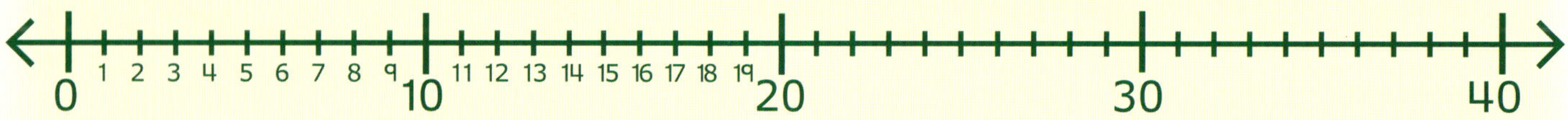

Locate 31 on the number line by tracing the path from 0 to 31.

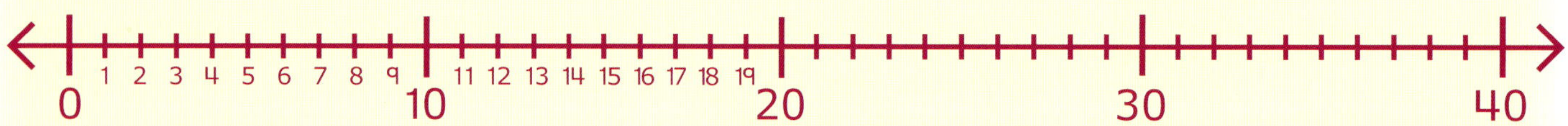

Locate 25 on the number line by tracing the path from 0 to 25.

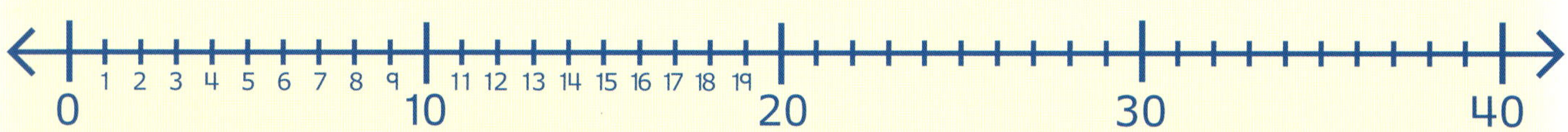

Locate 18 on the number line by tracing the path from 0 to 18.

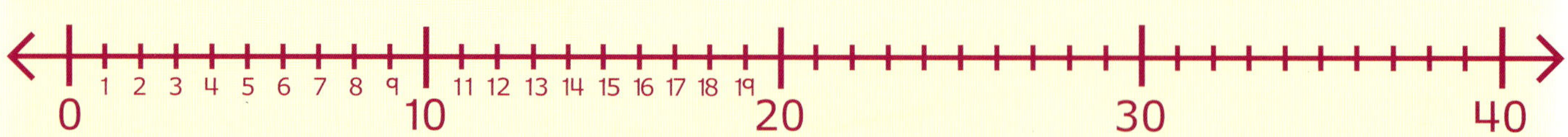

Locate 30 on the number line by tracing the path from 0 to 30.

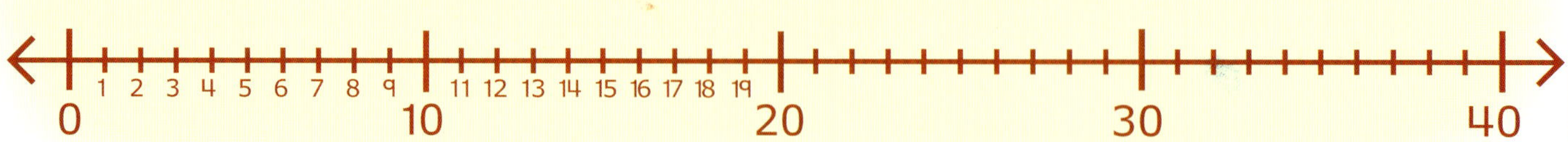

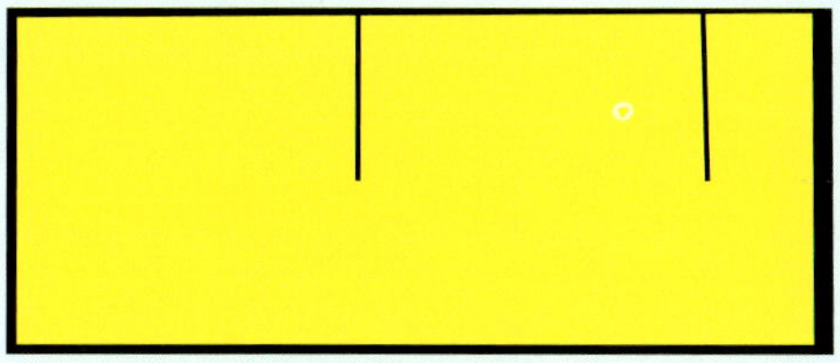

Here is a section of a 2 inch ruler. Number both inches on the 2 inch ruler.

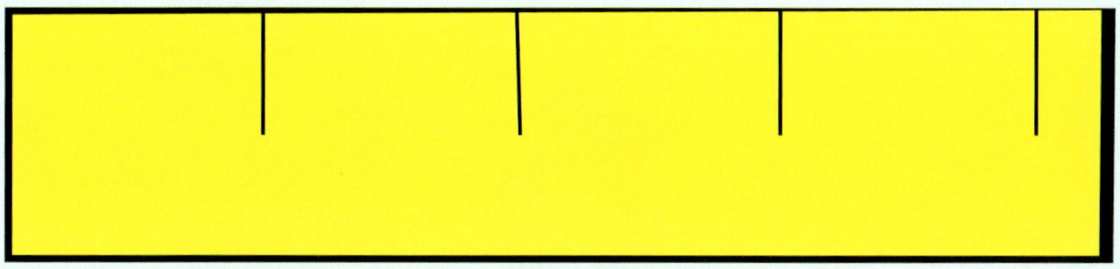

Here is a 4 inch ruler. Number the inches on the ruler.

Estimate the length of each fish below.

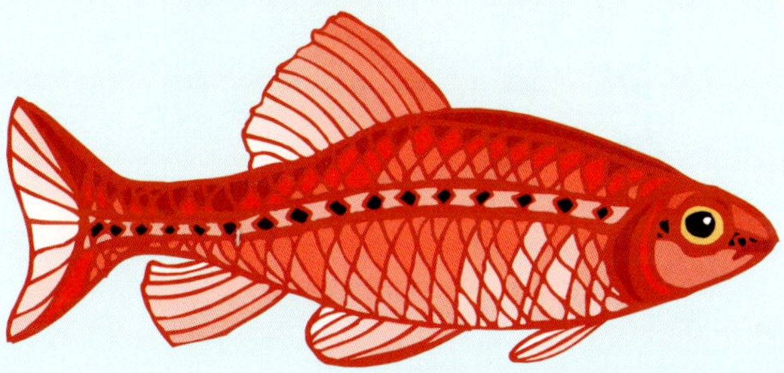

____

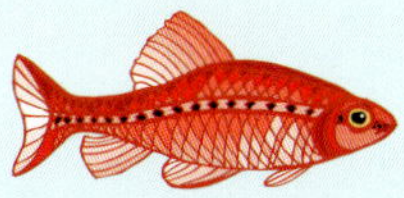

____

____

__________ __________ __________ __________

Use the clues and the measuring stick to find each person's name.

Bob is shorter than Kim. Dan has curly hair and is taller than Lee.

Use your completed diagram to write in the answers below.

Bob is __________ feet tall.

Bob and __________ are shorter than Dan.

__________ is 1 foot taller than Dan.

Say and write the sum of each problem.

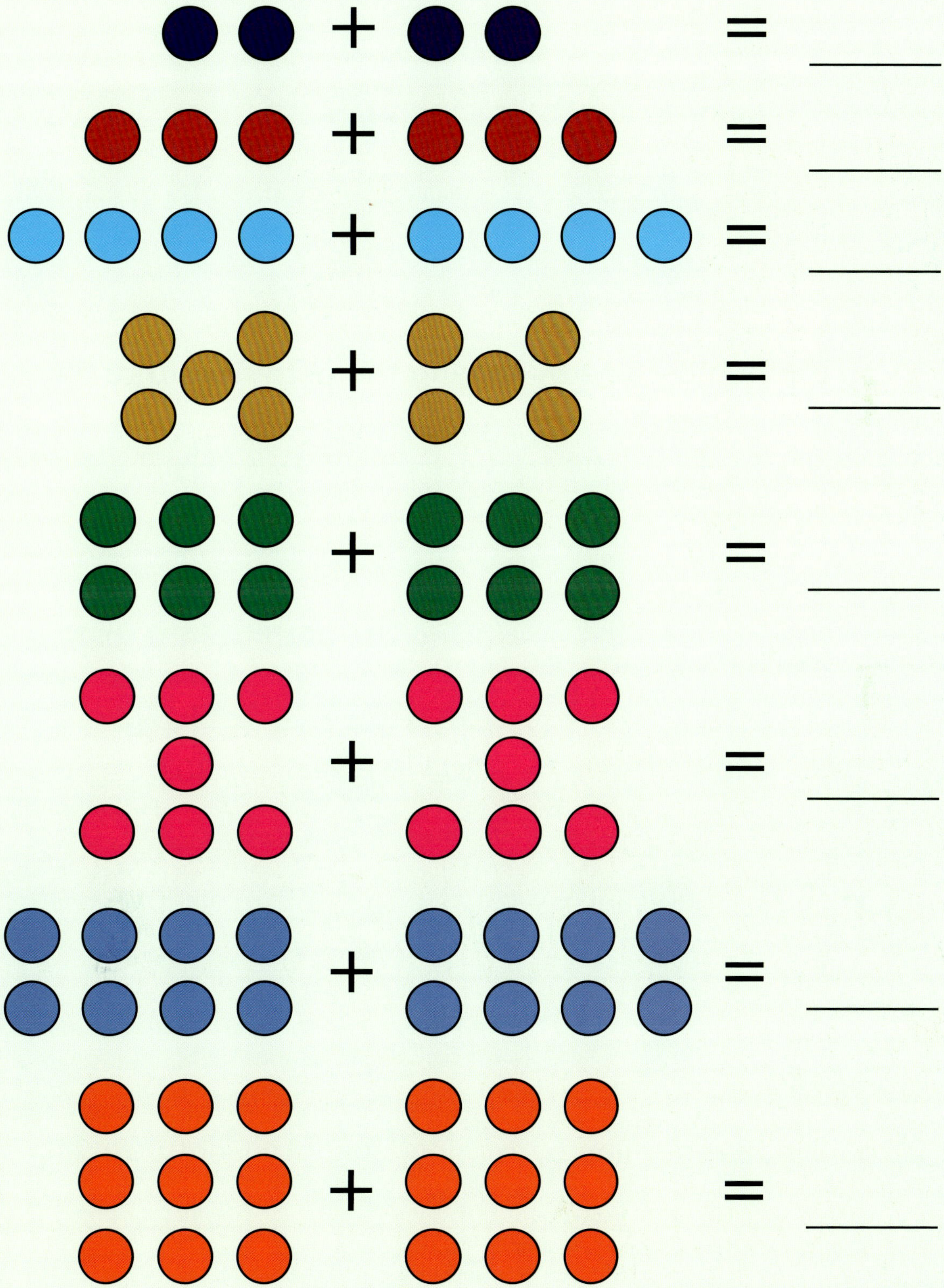

Locate 18 on the number line by tracing the path from 0 to 18.

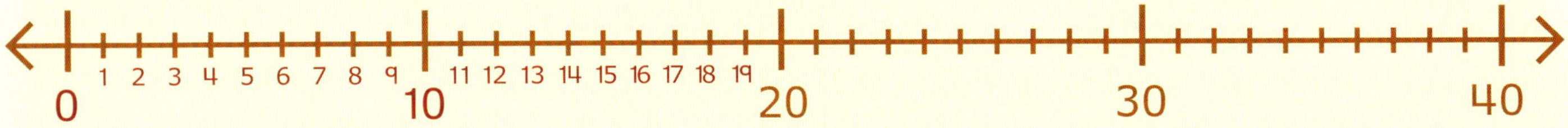

Locate 25 on the number line by tracing the path from 0 to 25.

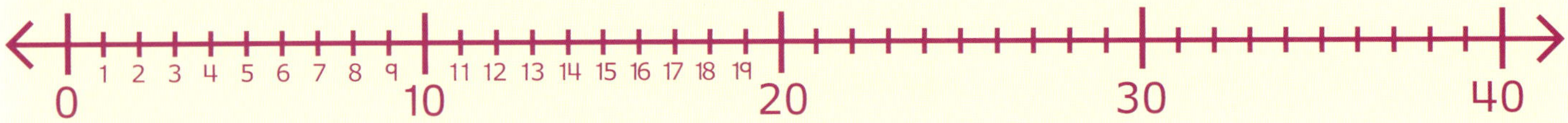

Locate 34 on the number line by tracing the path from 0 to 34.

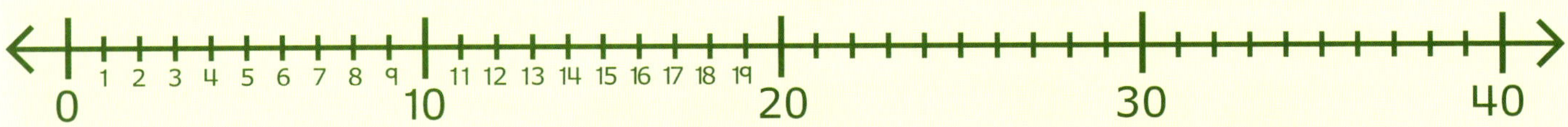

Locate 11 on the number line by tracing the path from 0 to 11.

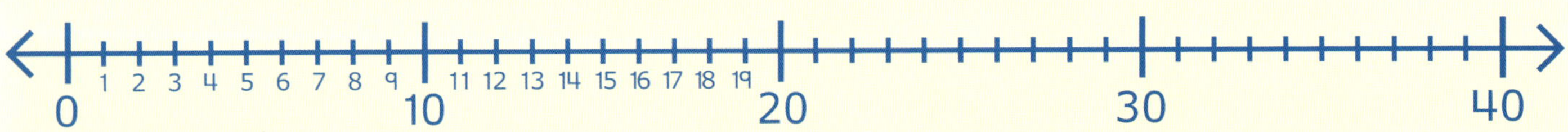

Locate 15 on the number line by tracing the path from 0 to 15.

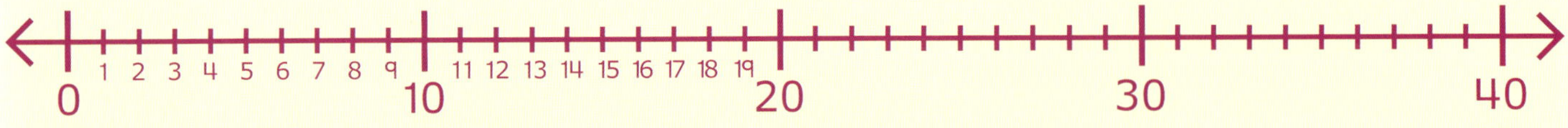

Locate 21 on the number line by tracing the path from 0 to 21.

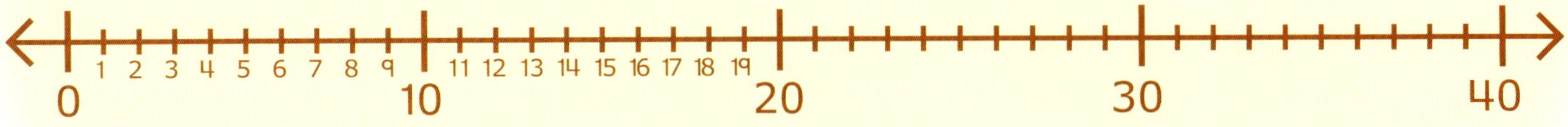

Fill in the missing numerals
to complete the pattern.

10	15		25			
		60	70			100
40	45		55			
30		50				90
55	60			75		
70	75			90		

Telling Time

The short hand (red) on the clock points to the hour. Write the hour below each clock.

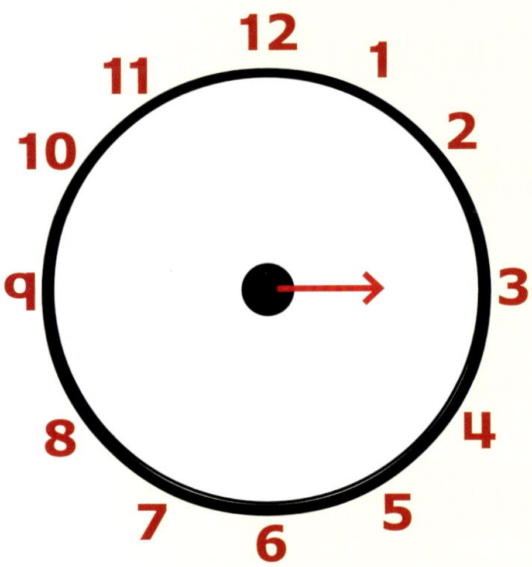

______

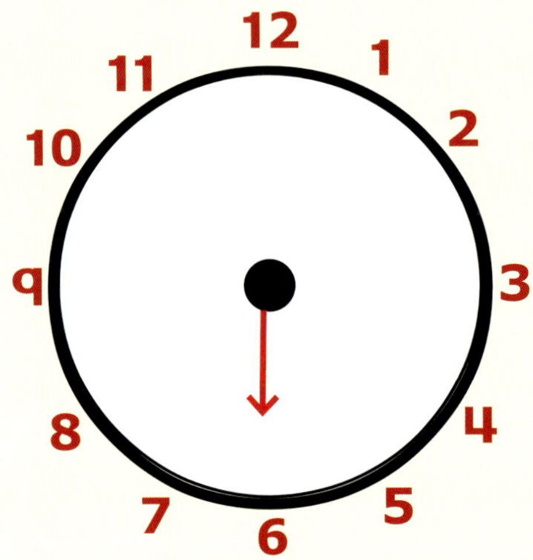

______

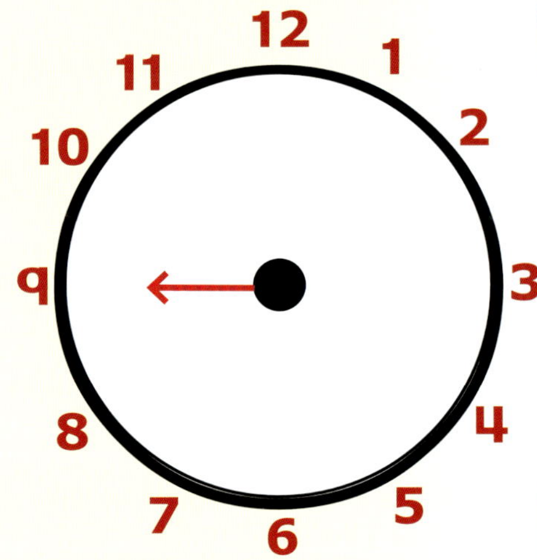

______

The long hand (blue) on the clock points to the minutes. There are 60 minutes in an hour. Write the number of minutes below the clock.

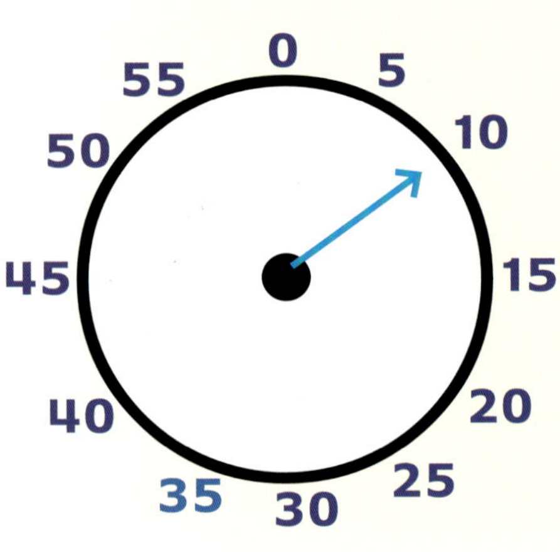

______

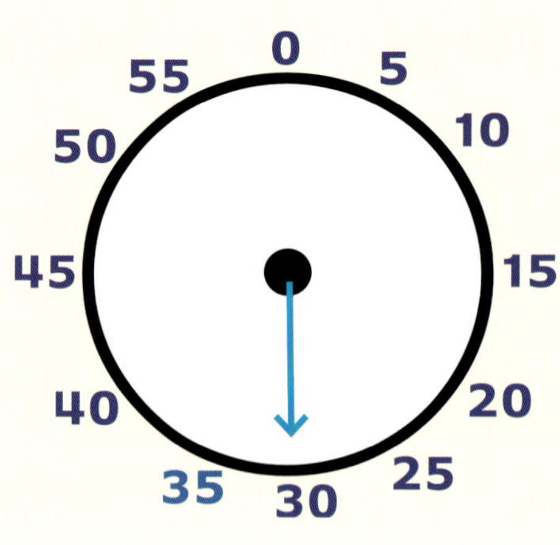

______

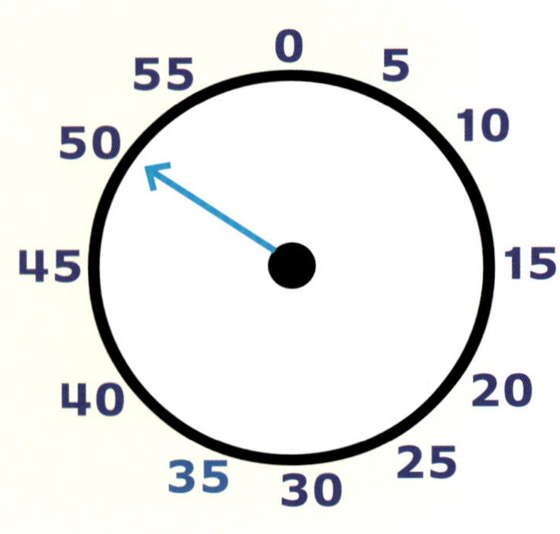

______

Draw line segments to connect each time to the correct clock.

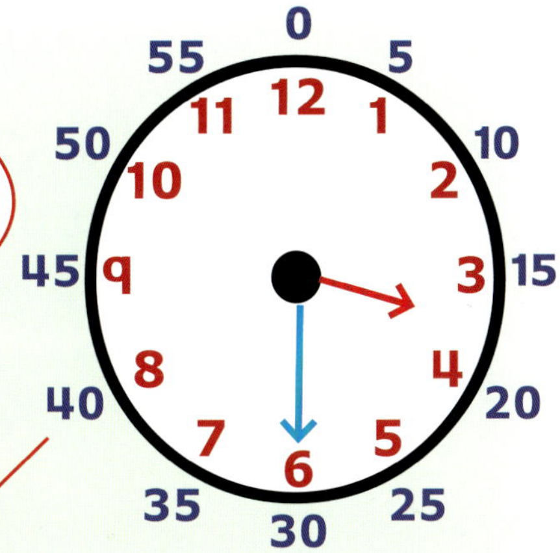

1 hour, 15 minutes

1:15

Example

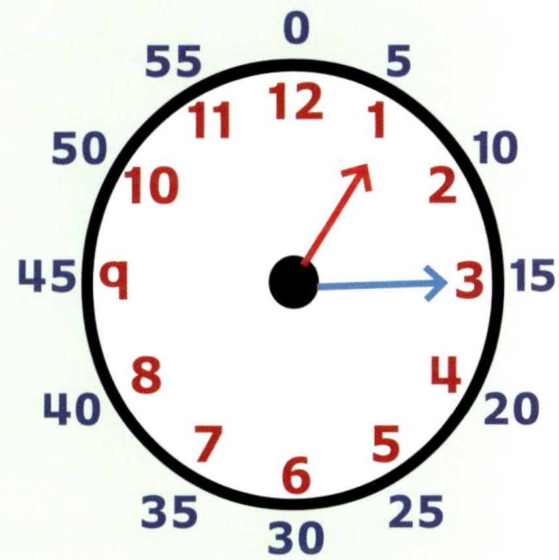

3 hours, 30 minutes

3:30

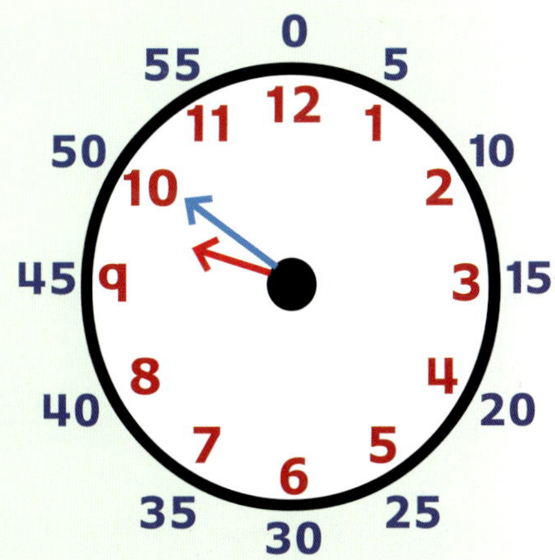

5 hours, 40 minutes

5:40

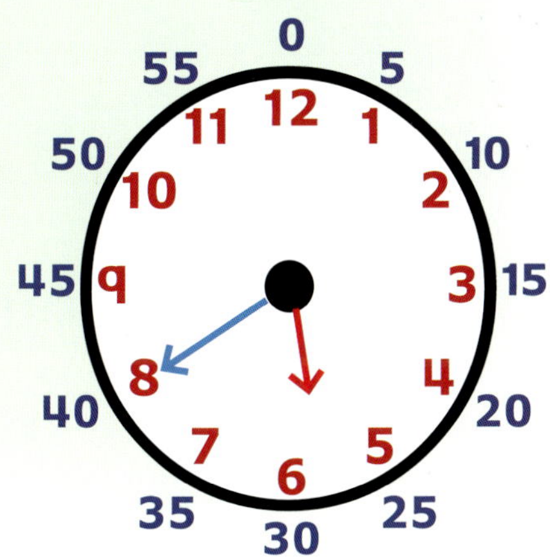

9 hours, 50 minutes

9:50

Draw line segments to connect each time to the correct clock.

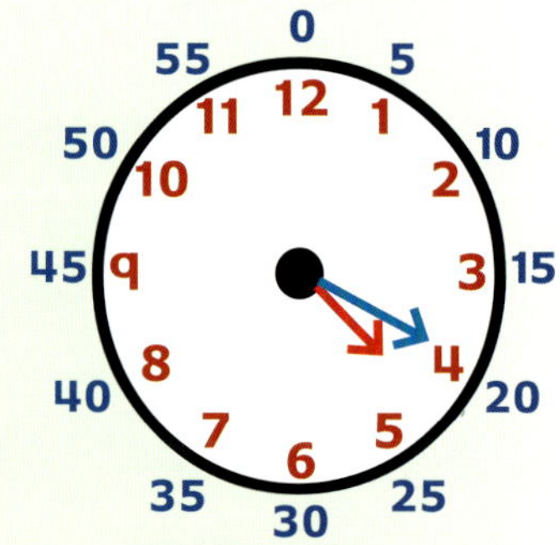

2:10

2 hours, 10 minutes

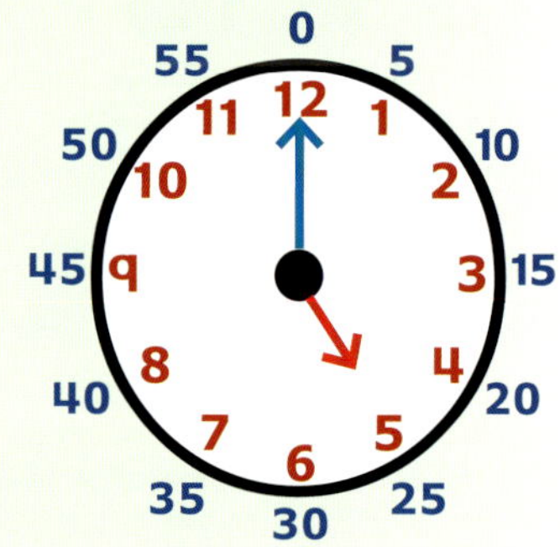

4:20

4 hours, 20 minutes

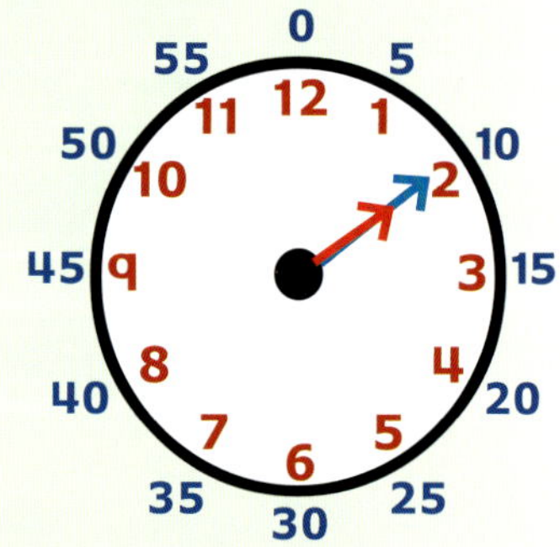

6:00

6 hours, 0 minutes

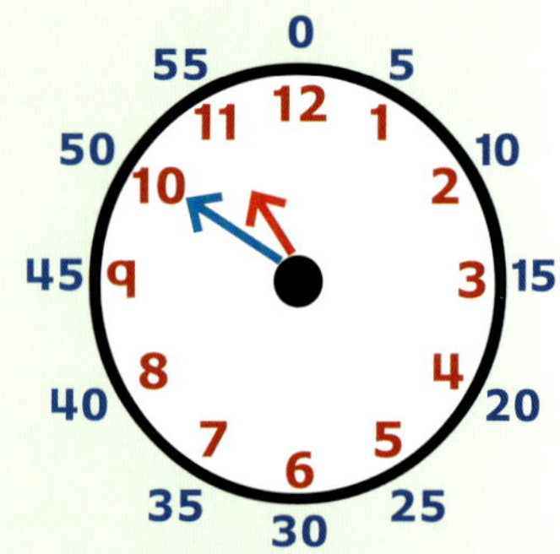

10:50

10 hours, 50 minutes

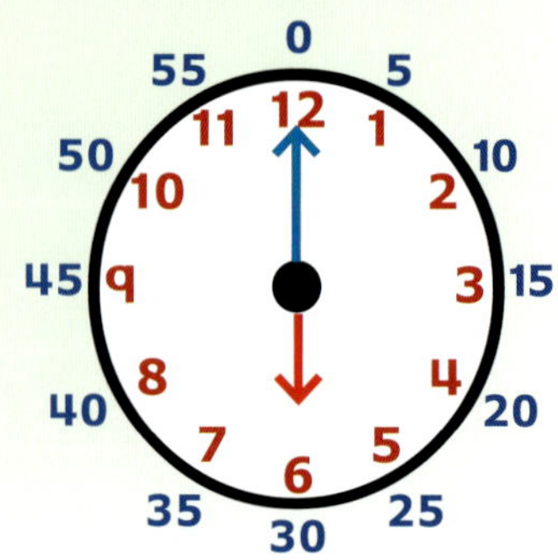

Draw each time on the clock.

1:10

1 hour, 10 minutes

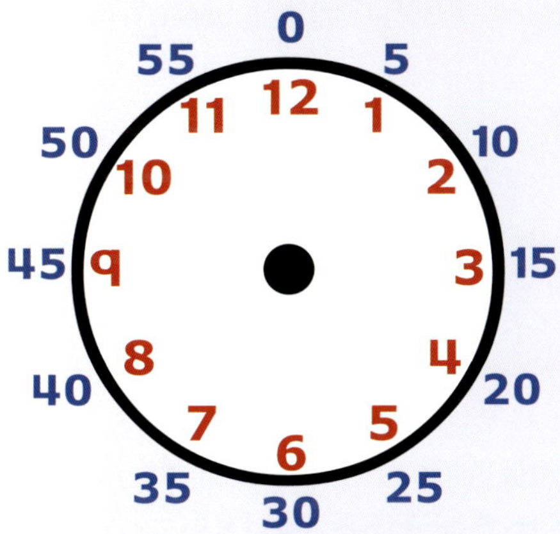

3:15

3 hours, 15 minutes

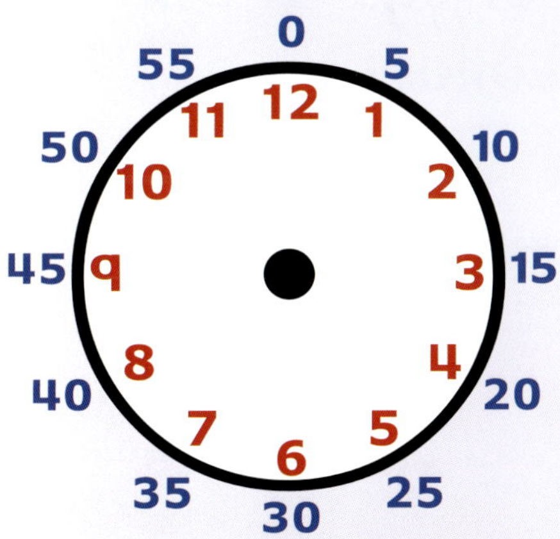

5:40

5 hours, 40 minutes

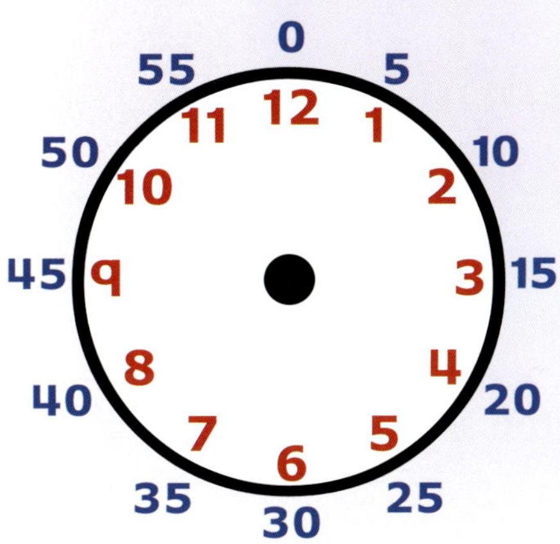

11:30

11 hours, 30 minutes

Answer the questions and explain your thinking.

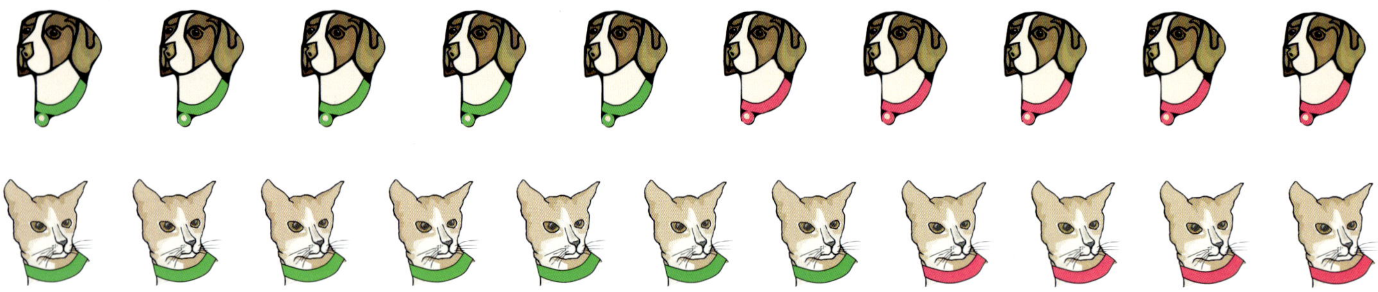

Which set has more members, the set of dogs or the set of cats?

Which set has less members, the set of cats with pink collars or the set of dogs with green collars?

Which set has more members, the set of pink collars or the set of green collars?

Which set has less members, the set of dogs with pink collars or the set of cats with green collars?

Which two sets have the same number of members?

Say and circle the answers.

What time is it on the clock? Is this the time you would probably go to bed? Yes or no?

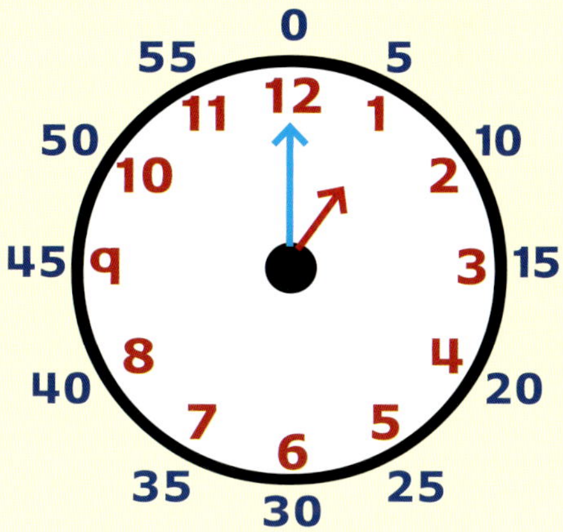

What time is it on the clock? Is this the time you would probably eat supper? Yes or no?

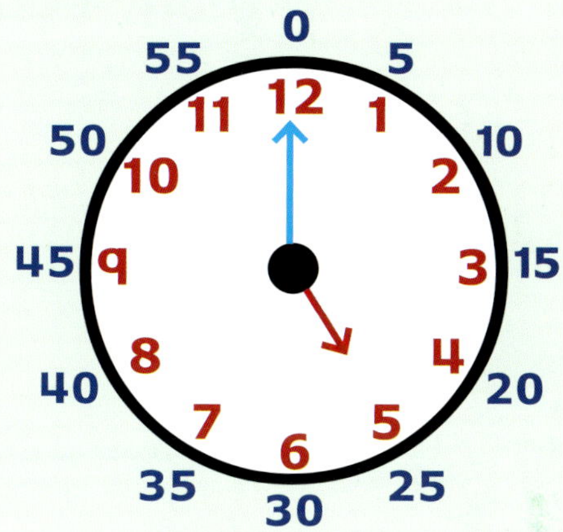

What time is it on the clock? Is this the time you would probably take a bath? Yes or no?

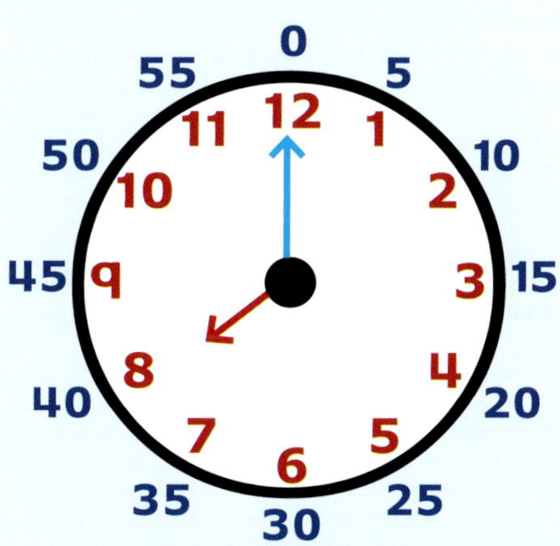

What time is it on the clock? Is this the time you would probably get up to start the day? Yes or no?

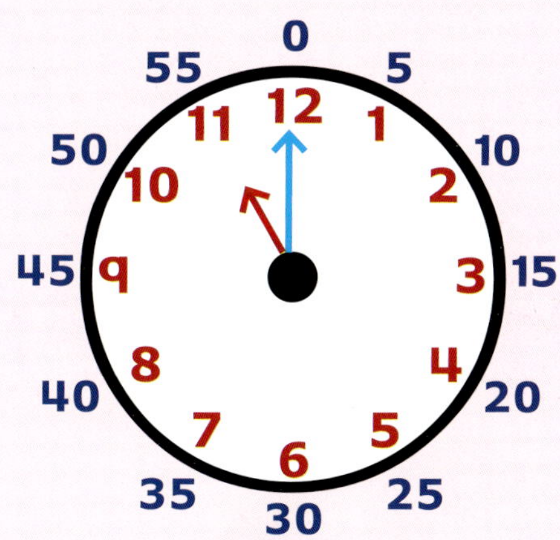

Write each number sentence.

Nine eggs are in an egg carton, then 6 eggs are used for breakfast. How many eggs are left in the carton?

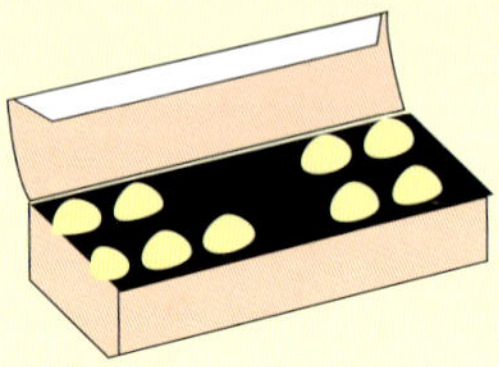

_______ - _______ = _______

Eight books are on a book shelf, then 6 books are returned to the library. How many books are left on the book shelf?

_______ - _______ = _______

Nine tomatoes are on a plate, then 4 tomatoes are used for a salad. How many tomatoes are left on the plate?

_______ - _______ = _______

Show the Time

Show the time you usually eat dinner.

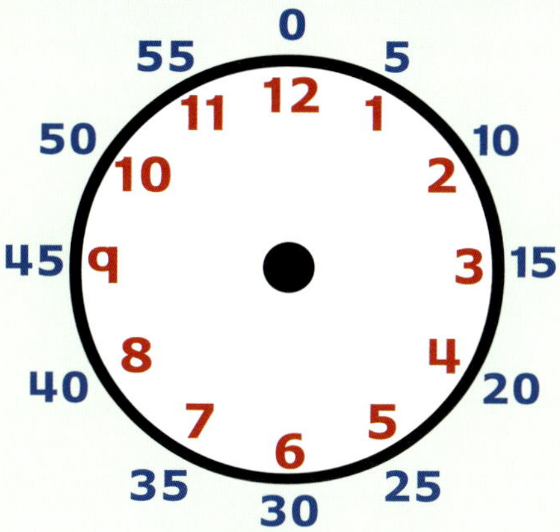

Show the time you usually go to bed.

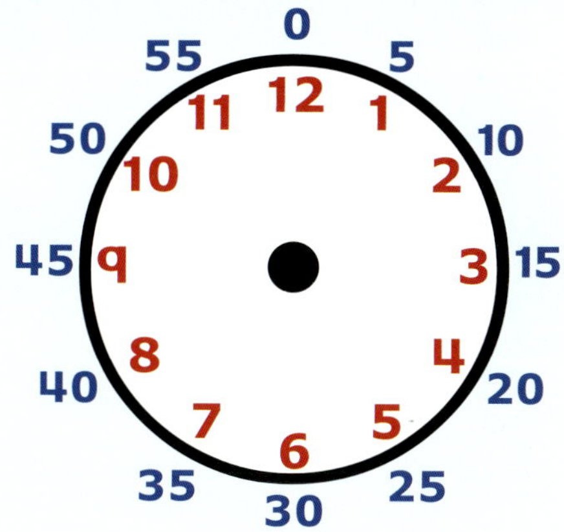

Show the time you usually wake up.

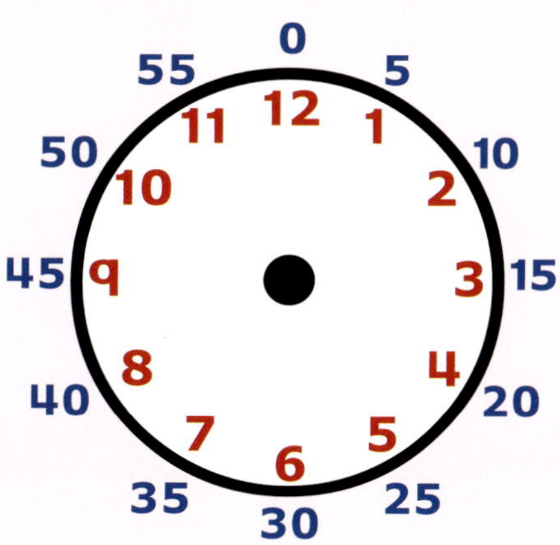

Show the time you usually eat lunch.

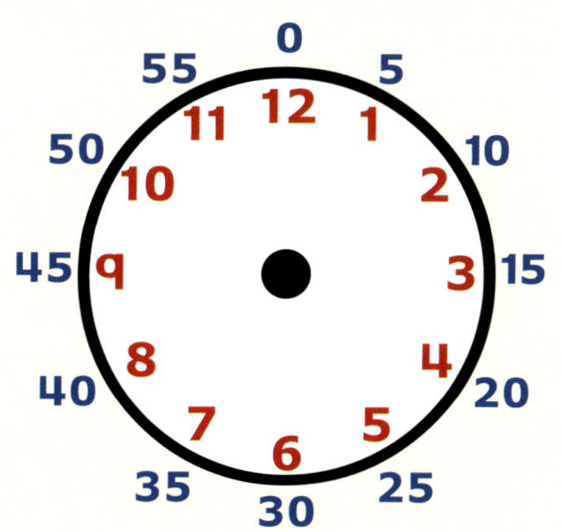

Write each number sentence.

Eight books are on a book shelf, then 5 books are returned to the library. How many books are left on the book shelf?

________ - ________ = ________

Nine tomatoes are on a plate, then all the tomatoes are used for a salad. How many tomatoes are left on the plate?

________ - ________ = ________

Nine eggs are in an egg carton, then 5 eggs are cooked. How many eggs are left in the carton?

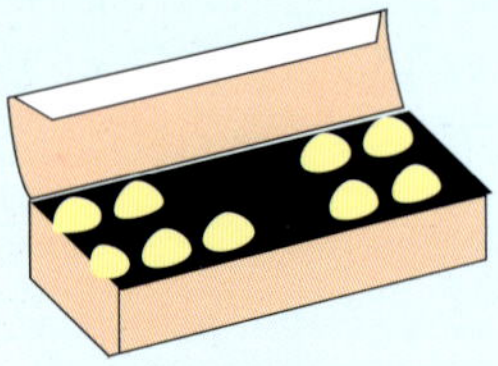

________ - ________ = ________

Use the bar graph to find the answers.

How many cars of each color were parked in the parking lot?

______ ______ ______

How many birthdays were in each month?

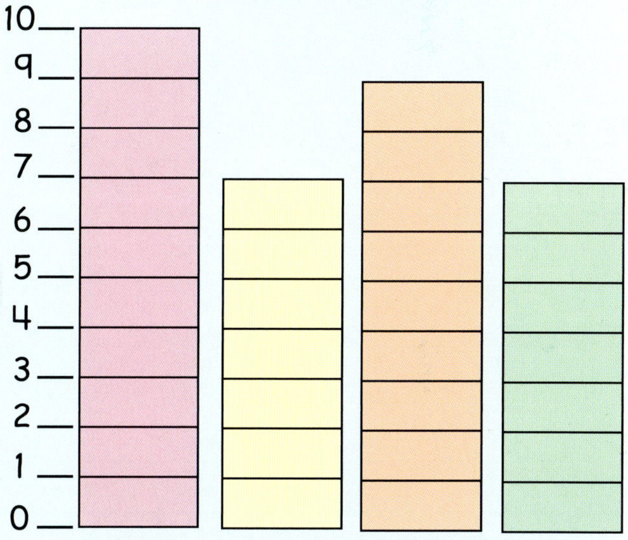

______ April ______ May ______ June ______ July

How many chocolates were eaten by each child?

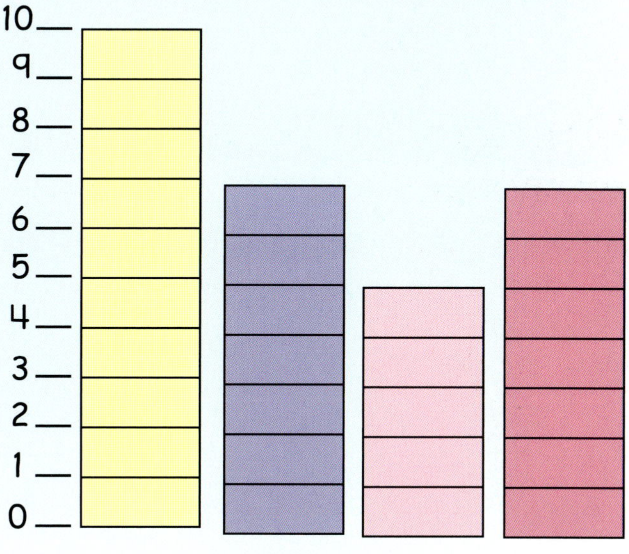

______ Peggy ______ Billy ______ Mary ______ Sam

Locate each numeral on the number line by tracing the path from 0.

40

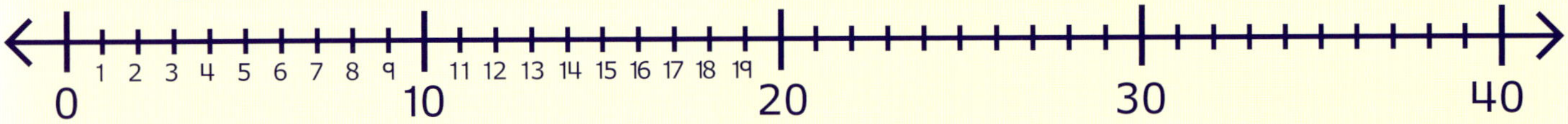

25

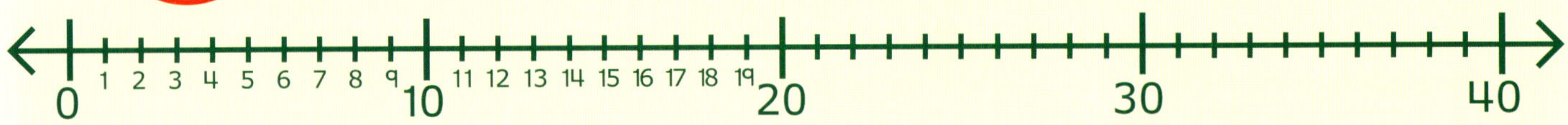

19

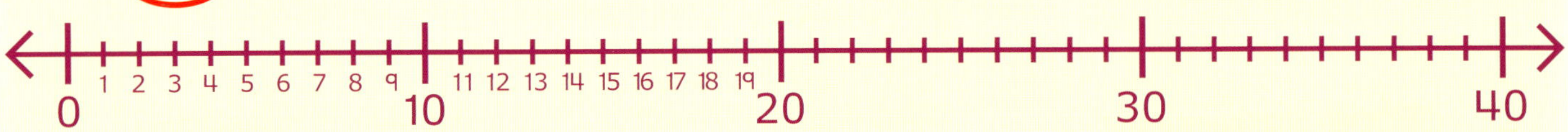

31

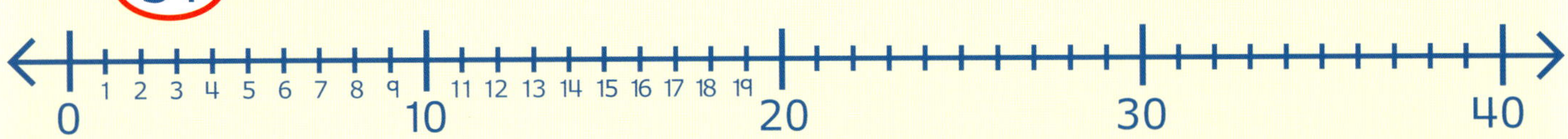

27

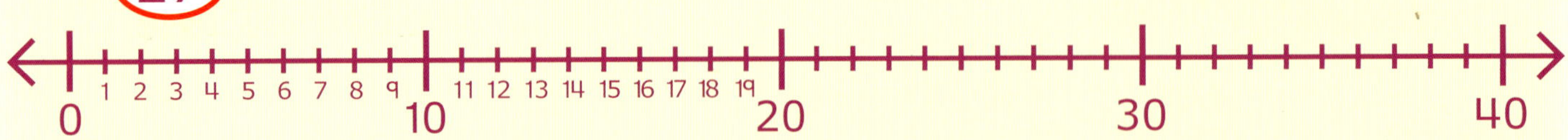

35

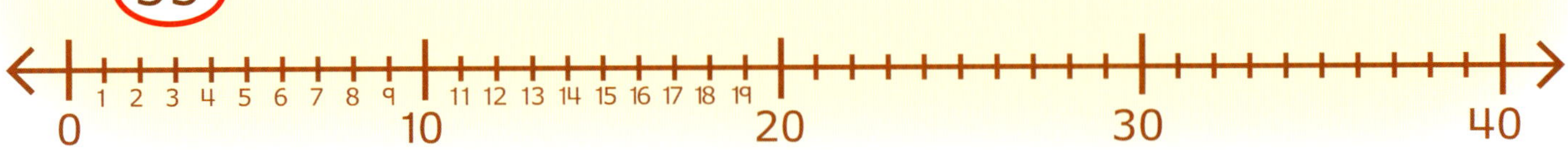

Circle the flat pattern that can be made into a shape like a soup can. Put an X over any shape that cannot. Explain each of your answers.

Do the problems. Go in order from left to right. Then, on the next page, connect the answer dots to find the mystery animal.

$\begin{array}{r} 10 \\ +10 \\ \hline \end{array}$	$\begin{array}{r} 20 \\ +40 \\ \hline \end{array}$	$\begin{array}{r} 30 \\ +40 \\ \hline \end{array}$	$\begin{array}{r} 10 \\ +20 \\ \hline \end{array}$
$\begin{array}{r} 10 \\ +50 \\ \hline \end{array}$	$\begin{array}{r} 40 \\ +20 \\ \hline \end{array}$	$\begin{array}{r} 30 \\ +20 \\ \hline \end{array}$	$\begin{array}{r} 20 \\ +10 \\ \hline \end{array}$
$\begin{array}{r} 10 \\ +30 \\ \hline \end{array}$	$\begin{array}{r} 30 \\ +20 \\ \hline \end{array}$	$\begin{array}{r} 40 \\ +30 \\ \hline \end{array}$	$\begin{array}{r} 30 \\ +40 \\ \hline \end{array}$
$\begin{array}{r} 20 \\ +30 \\ \hline \end{array}$	$\begin{array}{r} 40 \\ +60 \\ \hline \end{array}$	$\begin{array}{r} 30 \\ +30 \\ \hline \end{array}$	$\begin{array}{r} 30 \\ +50 \\ \hline \end{array}$
$\begin{array}{r} 10 \\ +10 \\ \hline \end{array}$	$\begin{array}{r} 40 \\ +0 \\ \hline \end{array}$	$\begin{array}{r} 30 \\ +30 \\ \hline \end{array}$	$\begin{array}{r} 0 \\ +20 \\ \hline \end{array}$
$\begin{array}{r} 50 \\ +50 \\ \hline \end{array}$	$\begin{array}{r} 10 \\ +40 \\ \hline \end{array}$	$\begin{array}{r} 0 \\ +10 \\ \hline \end{array}$	$\begin{array}{r} 0 \\ +30 \\ \hline \end{array}$

I live on a farm with cows and chickens. You can make your clothes from my coat.

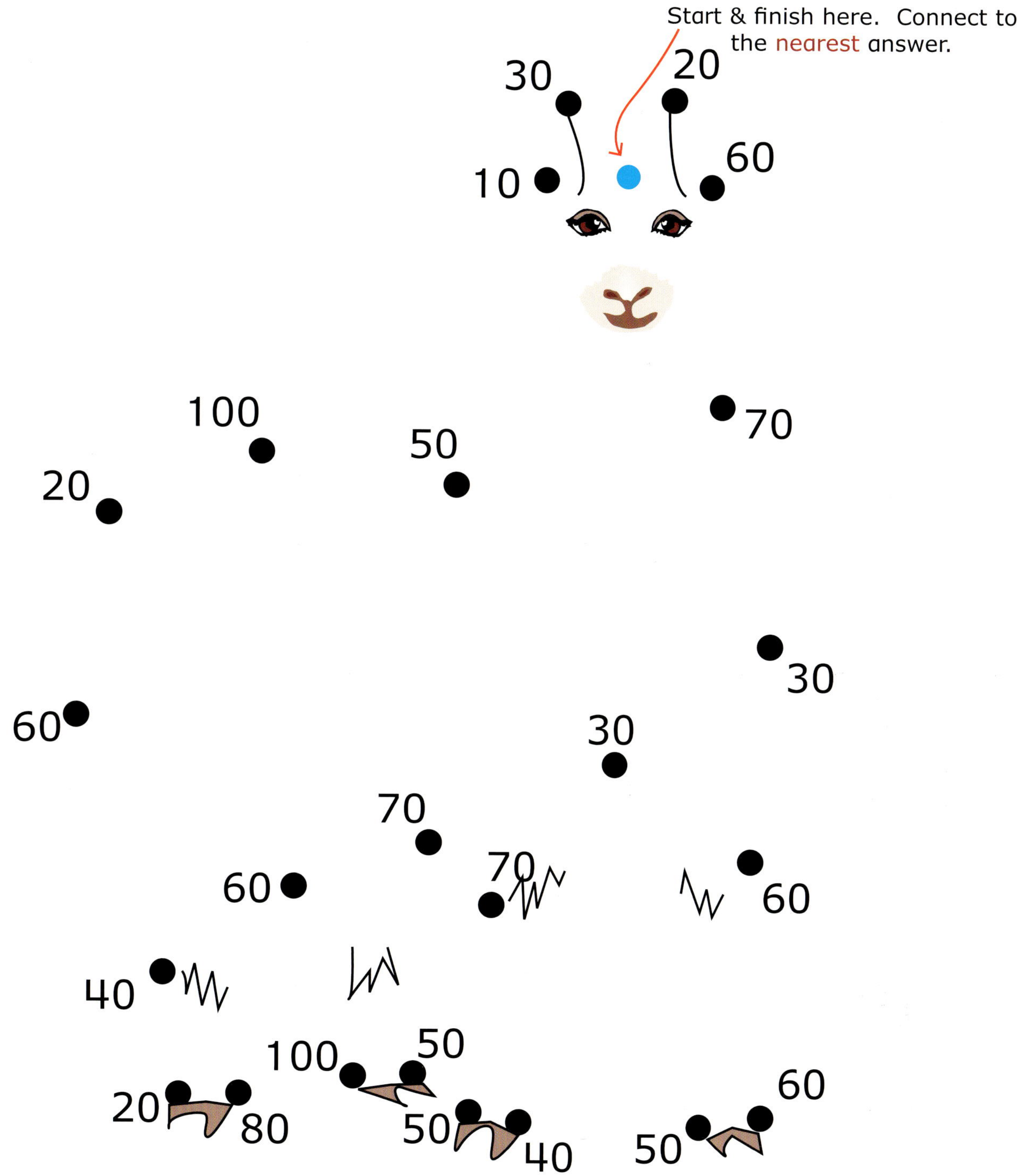

Color the mystery animal and add something new to the picture.

These boxes are all full of balls. All the balls are the same size. The smallest box holds 5 balls. Answer the questions and explain your thinking.

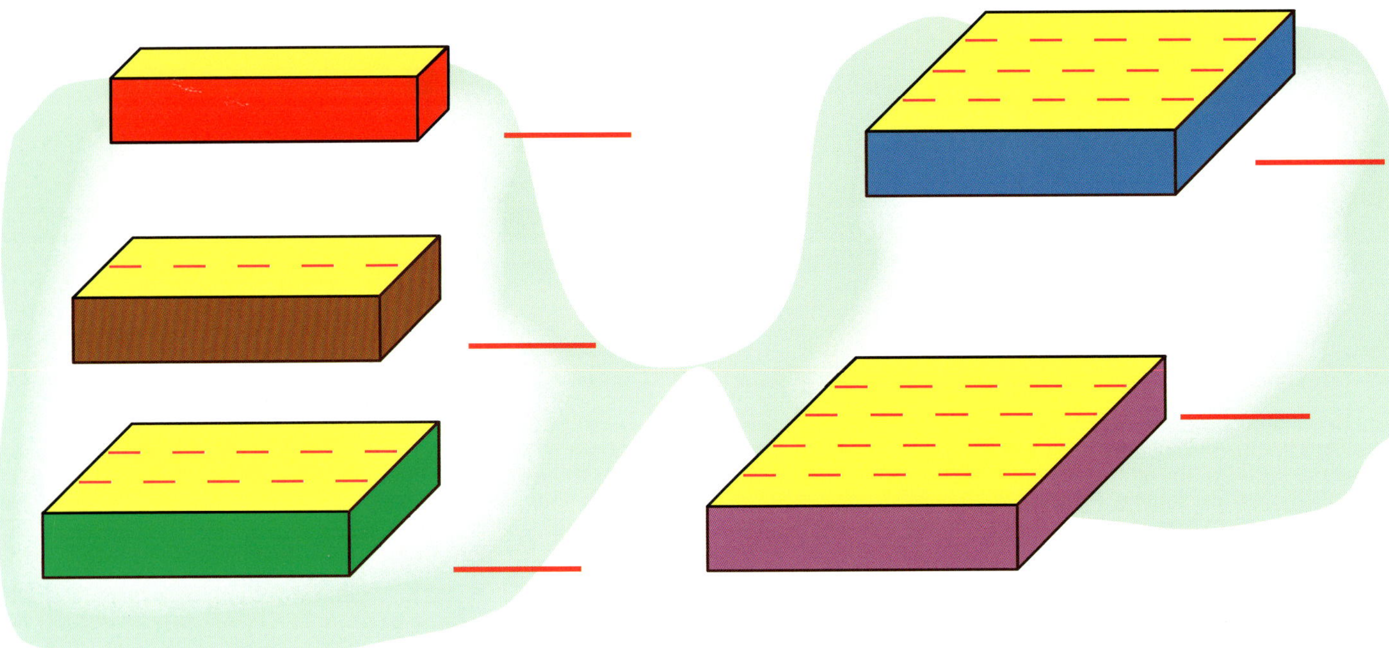

If the smallest box contains 5 balls, estimate the number of balls in the box below it.

__________ balls

Estimate the number of balls in the green box. __________ balls

Estimate the number of balls in the largest box. __________ balls

Write your estimate of balls in each box next to the boxes.

Create a bar graph that shows the teacher gave 9 notebooks and 7 pencils to the students.

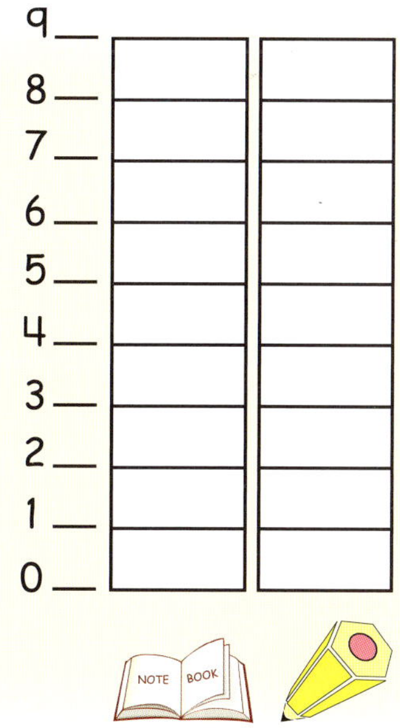

Create a bar graph that shows the classroom has 7 chairs, 9 desks, 1 stool, and 1 chalkboard.

Create a bar graph that shows there are 9 donuts, 8 brownies, 7 cookies, and 6 cakes left at the bakery.

1. Draw an X on the largest ball.
2. Circle the smallest ball.
3. Number the balls smallest to biggest using the numerals 1 to 5.

Create a bar graph that shows a pet store has 9 puppies and 9 kittens.

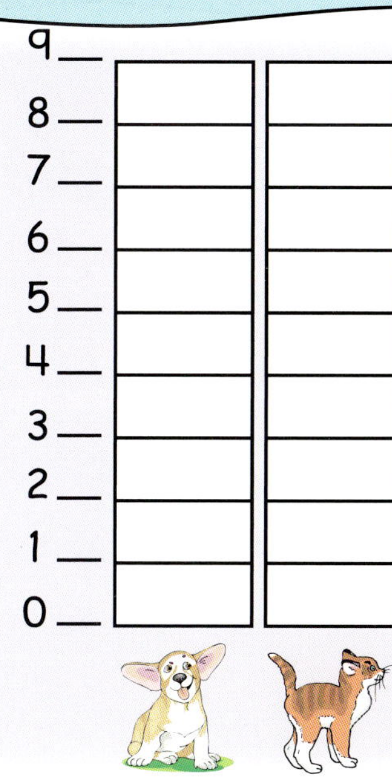

Create a graph that shows 9 pieces of chocolate, 8 red suckers, and 7 packs of gum.

Create a bar graph that shows 1 soda, 1 bag of jelly beans, 9 boxes of popcorn, and 7 caramel apples left at the movie snack bar.

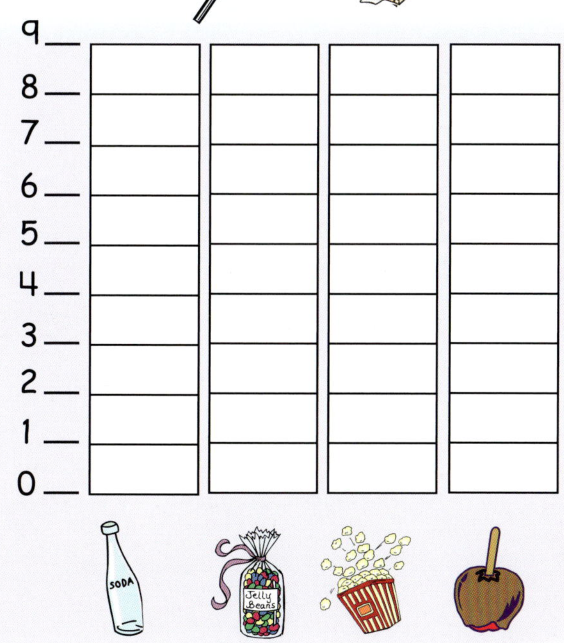

In each set, put an X on the container that holds the most. Circle the container that holds the least.

Example

You have 8 red balls and 1 yellow ball. If you put them in a box, closed your eyes, and picked only one ball, what color would it probably be? Explain your answer.

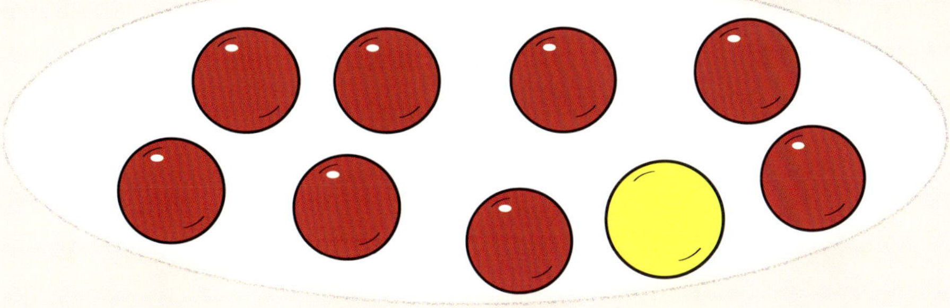

You have 3 purple balls and 10 red balls. If you put them in a box, closed your eyes, and picked only one ball, what color would it probably be? Explain your answer.

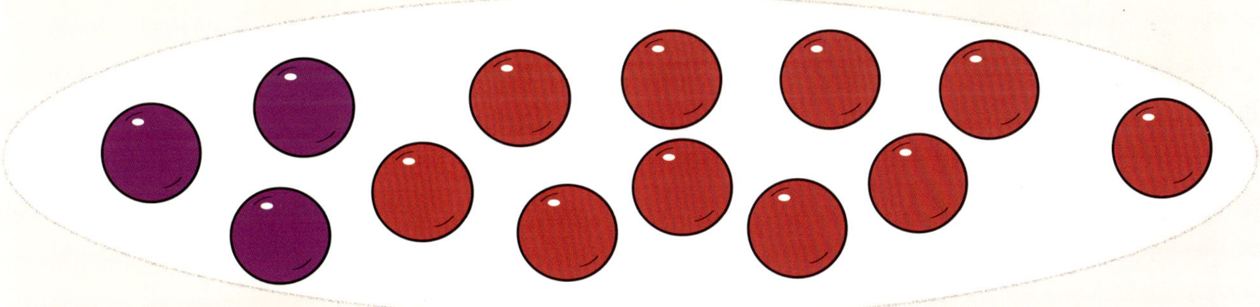

You have 12 orange balls and 4 black balls. If you put them in a box, closed your eyes, and picked only one ball, what color would it probably be? Explain your answer.

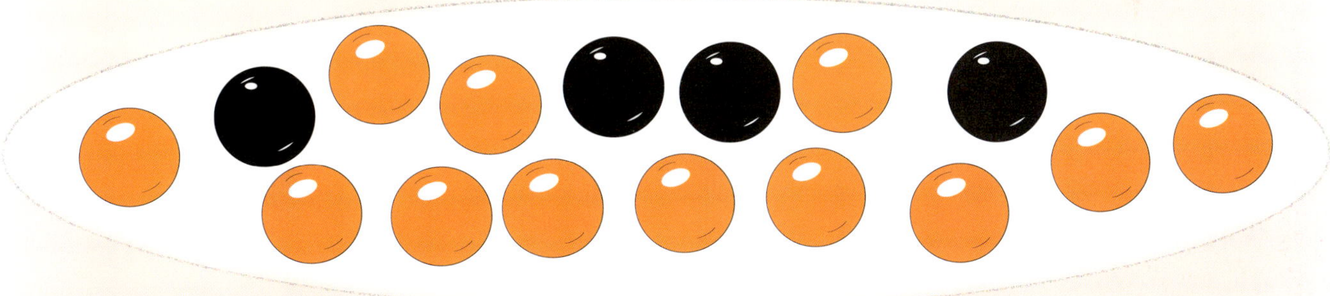

You have 7 red balls and 7 blue balls. If you put them in a box, closed your eyes, and picked only one ball, what color would it probably be? Explain your answer.

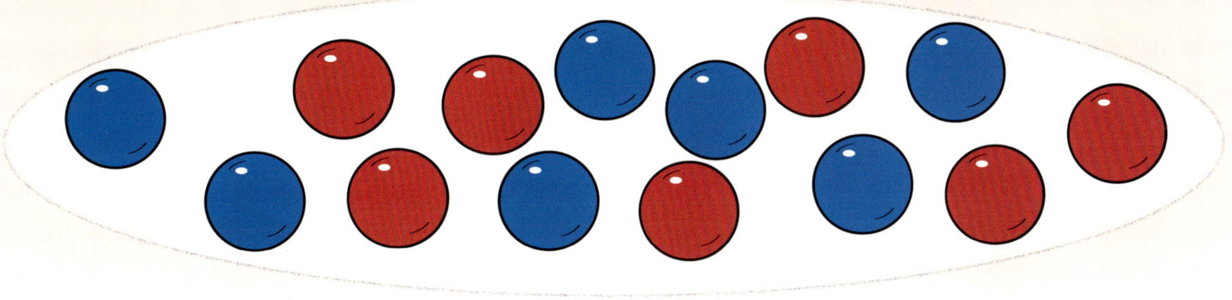

Trace and write the missing numerals, then count aloud by tens.

10 20 30 40 50

20 ______ 40 50 60

30 ______ 50 ______ 70 ______

40 50 ______ 70 ______ 90

50 ______ 70 80 ______ 100

Match the base ten blocks amounts with a matching amount of coins.

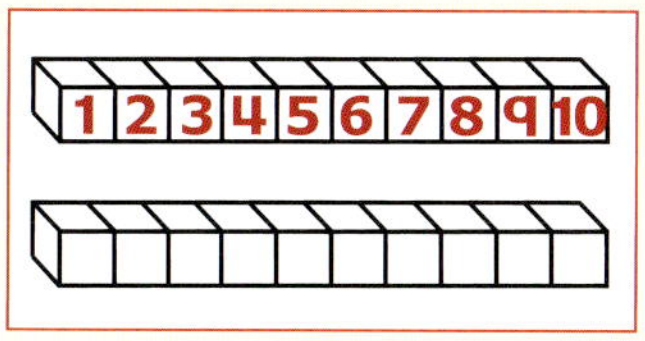

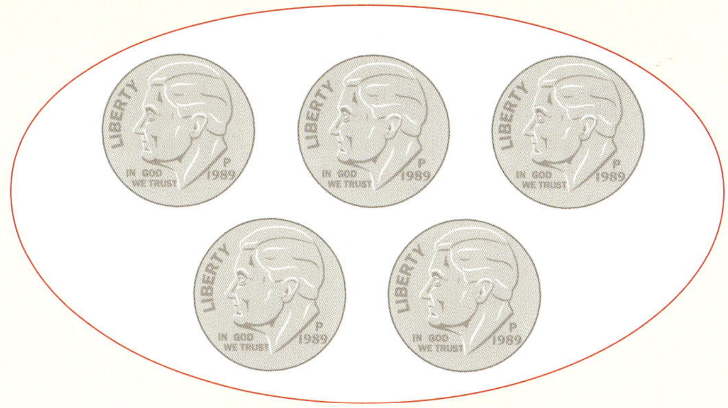

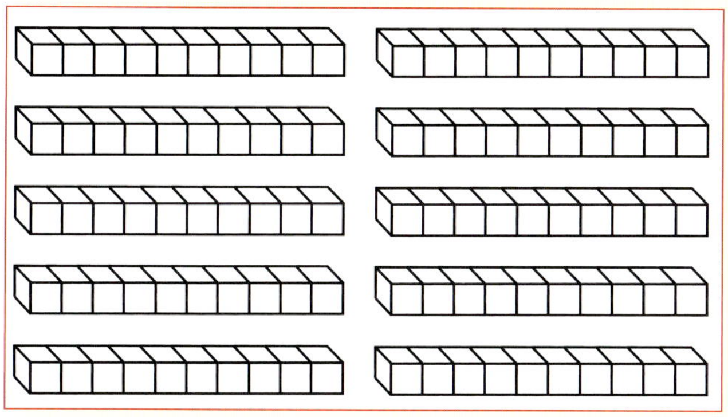

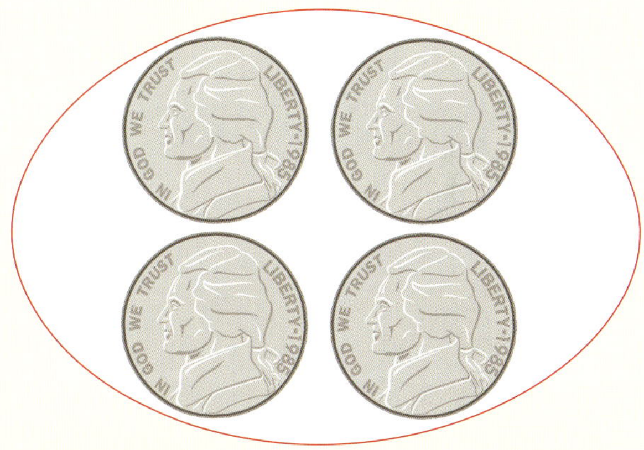

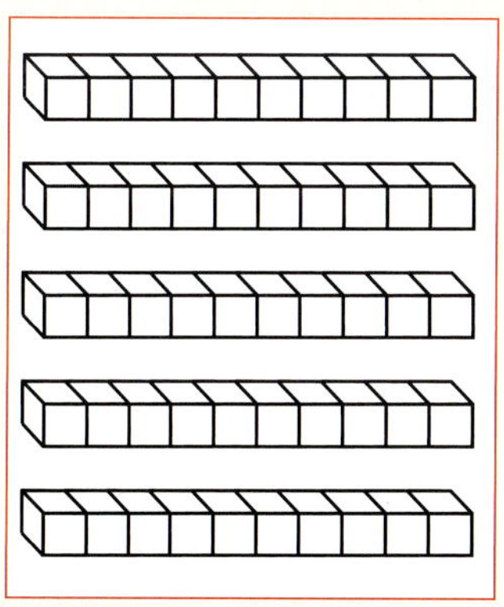

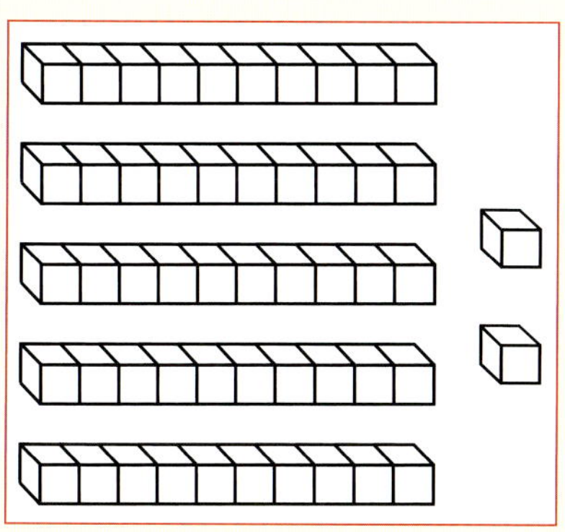

Circle the largest amount of base ten blocks. Put an X over the least amount of cents.

Do the problems. Go in order from left to right. Then, on the next page, connect the answer dots to find the mystery animal.

10 - 10	20 - 10	30 - 10	40 - 10
50 - 20	50 - 10	30 - 10	40 - 20
30 - 20	30 - 30	40 - 20	40 - 30
20 - 10	30 - 10	40 - 10	50 - 10
30 - 20	40 - 30	20 - 10	40 - 20
50 - 40	40 - 30	20 - 10	40 - 20

I don't brush my teeth, but I do take many baths. I really like to build dams.

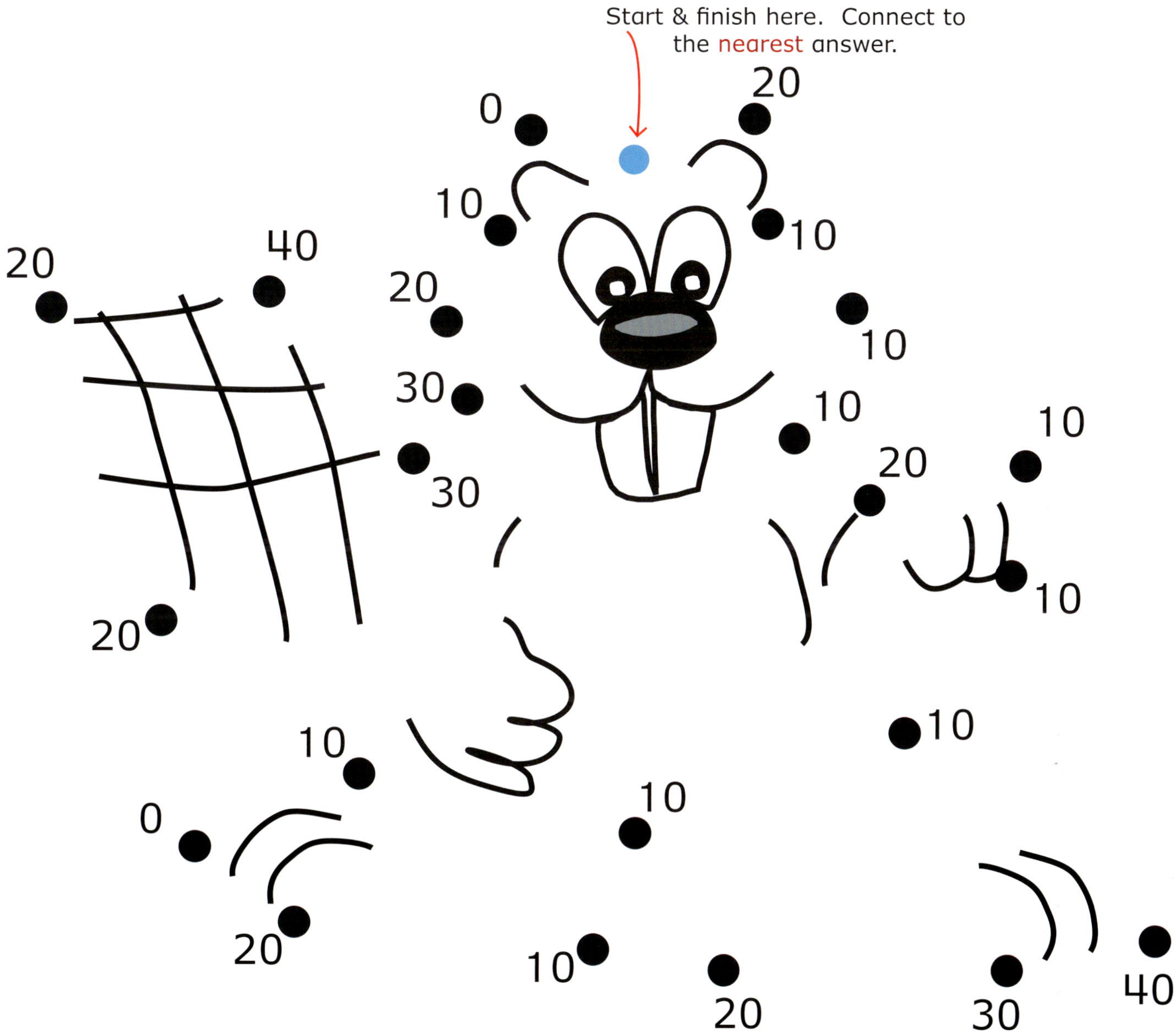

Color the mystery animal and add something new to the picture.

You have 9 red balls. If you put them in a box, closed your eyes, and picked only one ball, what color would it be? Explain your answer.

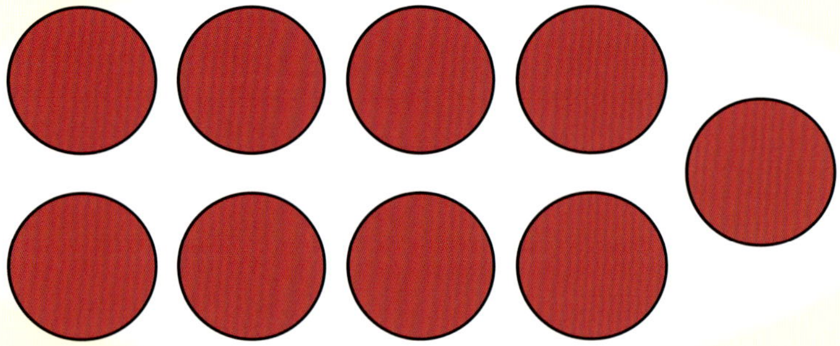

You have 8 red balls and 1 yellow ball. If you put them in a box, closed your eyes, and picked only one ball, what color would it probably be? Explain your answer.

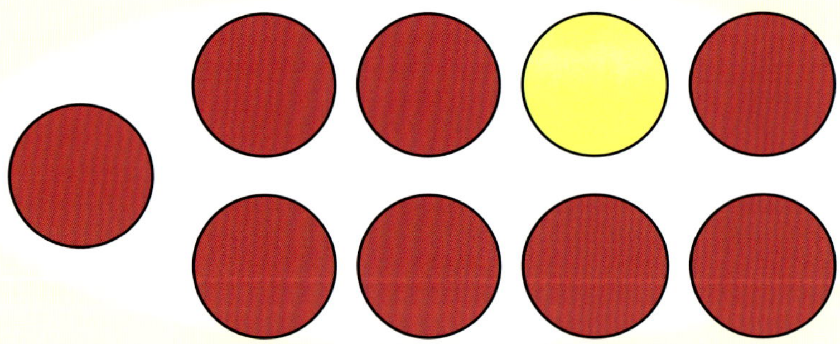

You have 6 green balls and 3 purple balls. If you put them in a box, closed your eyes, and picked only one ball, what color would it probably be? Explain your answer.

You have 3 orange balls and 4 brown balls. If you put them in a box, closed your eyes, and picked only one ball, what color would it probably be? Explain your answer.

You have 7 red balls and 1 yellow ball. If you put them in an empty box, closed your eyes, and picked only one ball, could you get a blue ball? Explain your answer.

Circle the correct answer.

1. If most of the cars in the tunnel are green, but some are blue, are there more red cars or green cars?

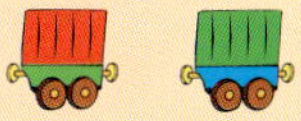

2. If the train cars in the tunnel look like this,

what color are most of the cars?

3. If the cars in the tunnel look like this,

what color are most of the cars?

4. If the cars in the tunnel look like this,

what color are most of the cars?

Do the problems. Go in order from left to right. Then, on the next page, connect the answer dots to find the mystery animal.

80 +10	50 +40	30 +20	30 +50
50 +20	40 +50	60 +20	70 +10
10 +60	30 +50	60 +30	30 +60
70 +20	20 +70	50 +30	30 +50
30 +30	40 +40	20 +20	10 +10
70 +10	10 +50	60 +10	80 +30

I can have a lifespan of up to 30 years. I can weigh as much as 2000 pounds, and I sometimes wear shoes.

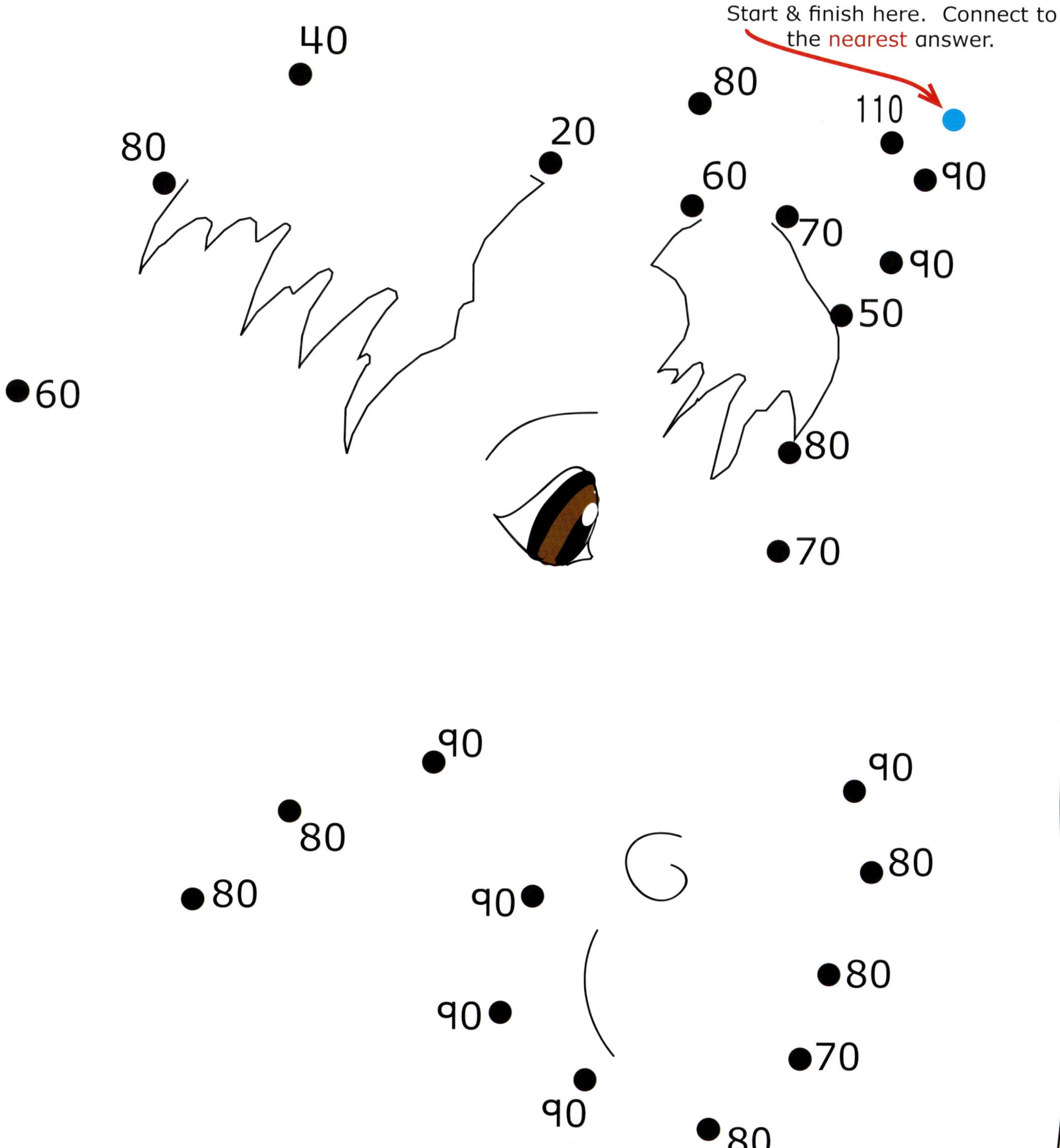

Color the mystery animal and add something new to the picture.

1	2	3	4	5	6	7	8	9	10
first	second	third	fourth	fifth	sixth	seventh	eighth	ninth	tenth

1 2 3 4 5 6 7 8 9 10

Write the order of each flower in the blank space.

Example

The orange flower is the first flower in the row.

The blue flower is the ____________ flower in the row.

The polka dot flower is the ____________ flower in the row.

The green flower is the ____________ flower in the row.

The pink flower is the ____________ flower in the row.

The purple flower is the ____________ flower in the row.

The rainbow flower is the ____________ flower in the row.

The red flower is the ____________ flower in the row.

The yellow flower is the ____________ in the row.

The white flower is the ____________ in the row.

Example

Circle the fifth computer mouse.

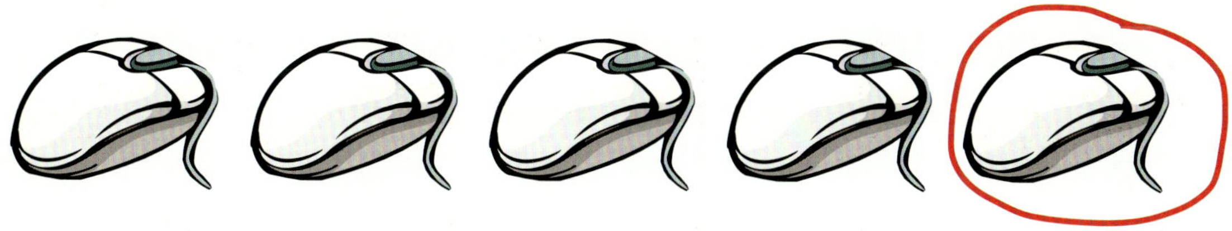

Circle the third mouse pad.

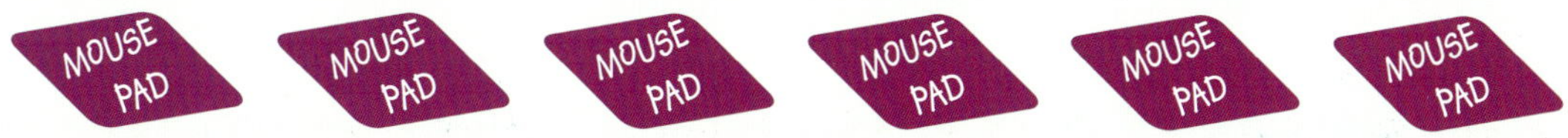

Circle the sixth computer monitor.

Circle the eighth CD.

Circle the seventh keyboard.

Circle the ninth pair of computer speakers.

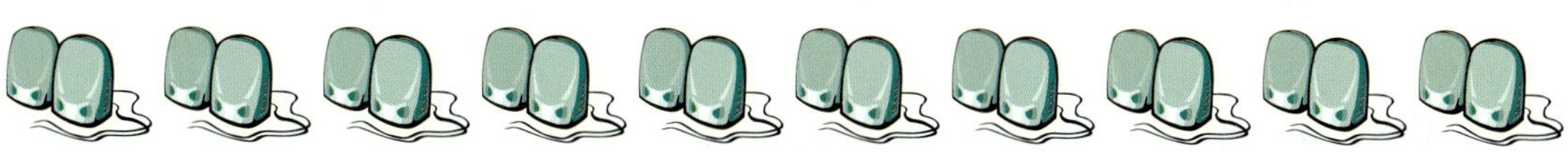

Circle the tenth cell phone.

Do the problems. Go in order from left to right. Then, on the next page, connect the answer dots to find the mystery object.

80 - 20   50 - 40   30 - 10   50 - 30

30 - 10   40 - 30   60 - 20   70 - 60

40 - 30   30 - 30   60 - 30   70 - 50

30 - 20   20 - 20   50 - 40   30 - 30

30 - 20   40 - 30   20 - 10   40 - 20

70 - 20   50 - 30   60 - 50   80 - 40

This sometimes has legs, but it does not walk. Mom and Dad often reward homework and chores with time watching this.

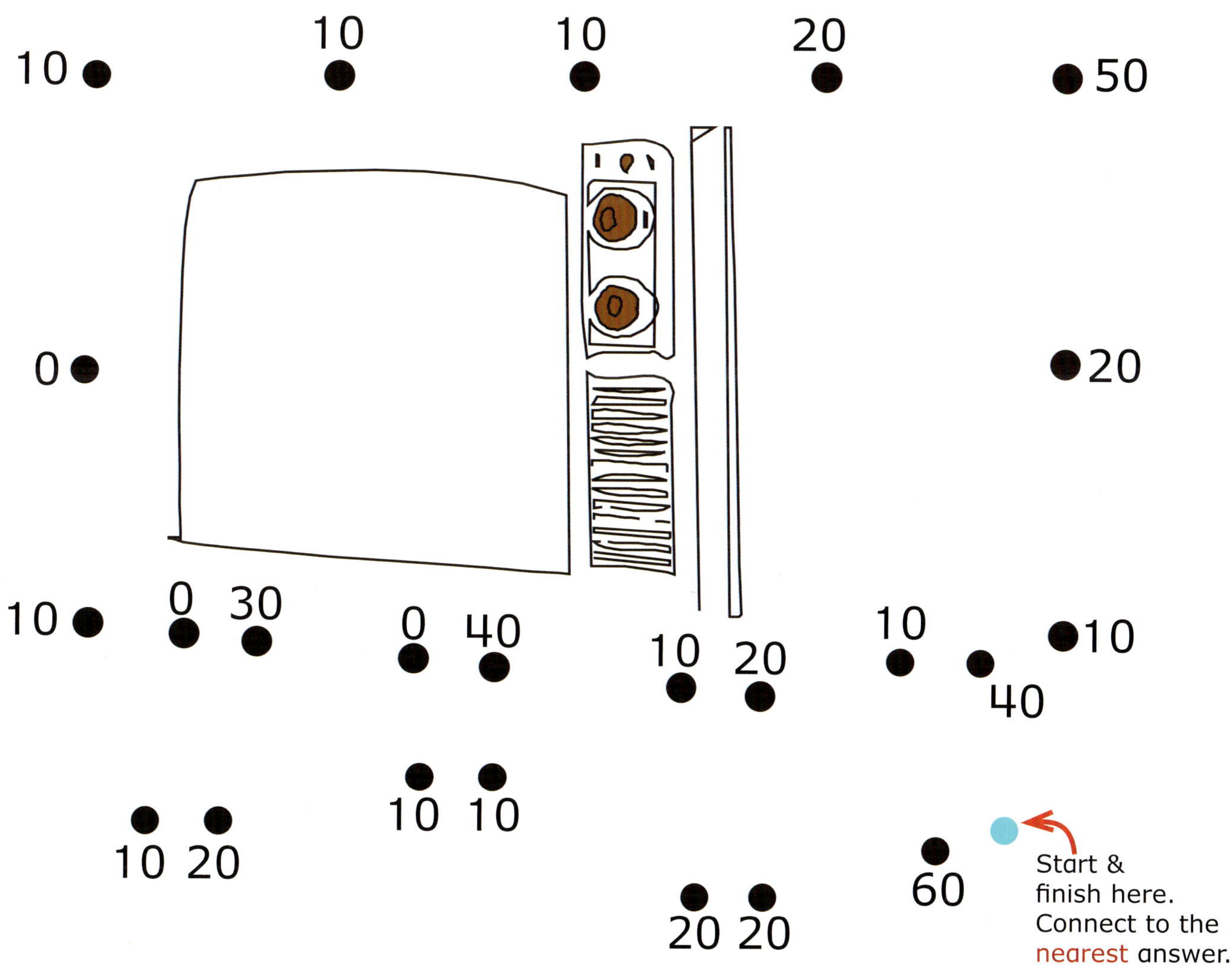

Color the mystery object and add something new to the picture.

Match each number sentence to the correct sum.

5 + 5 =	8
2 + 2 =	18
4 + 4 =	12
1 + 1 =	10
9 + 9 =	2
6 + 6 =	6
8 + 8 =	4
7 + 7 =	16
3 + 3 =	14

Example

Smallest to Largest Order

This box will hold 30 baseballs.

This box will hold 10 baseballs.

How many baseballs do you estimate this box will hold ?

____

Circle each set of numbers that makes 8.

EXAMPLE:

(8 + 0) 7 + 0 6 + 0 5 + 0 (7 + 1)

What makes 8?

4 + 4 3 + 5 6 + 1 1 + 7 2 + 6

What makes 9?

4 + 5 8 + 1 2 + 7 1 + 6 0 + 9

What makes 10?

2 + 8 4 + 6 5 + 5 8 + 1 7 + 3

What makes 11?

2 + 9 10 + 10 5 + 6 8 + 13 10 + 1

Circle each set of numbers that makes 12. 7 + 5

What makes 12?

2 + 10 11 + 0 9 + 3 6 + 6 1 + 11
11 + 4 2 + 9

What makes 13?

10 + 3 8 + 4 6 + 7 2 + 10 12 + 1

What makes 14?

7 + 7 9 + 5 8 + 6 0 + 14 13 + 1

What makes 15?

5 + 5 + 5 10 + 5 8 + 7 10 + 4 14 + 1

What makes 16?

8 + 8 9 + 9 12 + 4 10 + 6 1 + 15

Do the actions and count aloud, repeating each pattern five times.

TAP

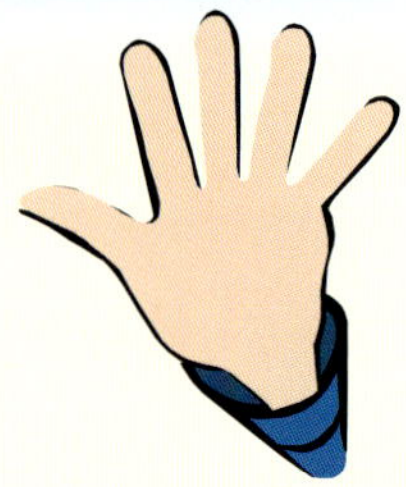

WAVE

STRETCH

BEND

Count the cubes in each set, then write the number sentence.

______ + ______ = ______

______ + ______ = ______

______ + ______ = ______

______ + ______ = ______

______ + ______ = ______

______ + ______ = ______

Circle each set of numbers that makes 17. 10 + 7

What makes 17?

3 + 14 9 + 8 16 + 1 15 + 2 0 + 17

What makes 18?

11 + 5 9 + 9 10 + 8 1 + 17 6 + 6 + 6

What makes 19?

1 + 9 1 + 18 20 + 1 10 + 9 17 + 2

What makes 20?

5 + 5 + 5 + 5 11 + 2 10 + 10 2 + 18

3 + 17 16 + 4

Count the cubes in each set, then write how many cubes there are all together.

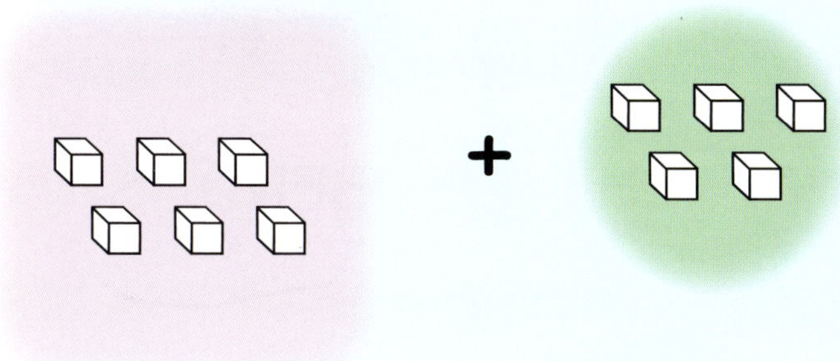

______ + ______ = ______

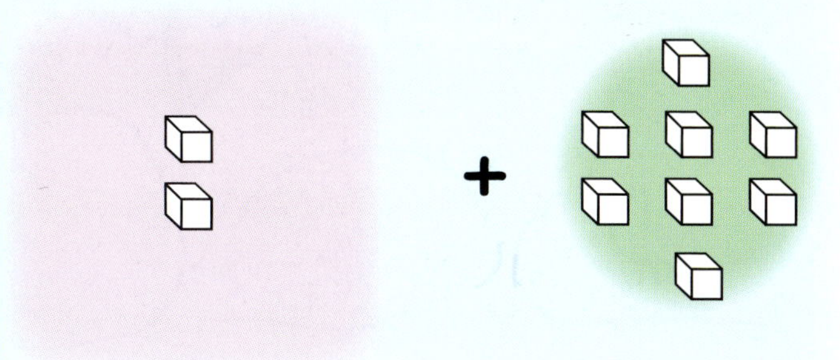

______ + ______ = ______

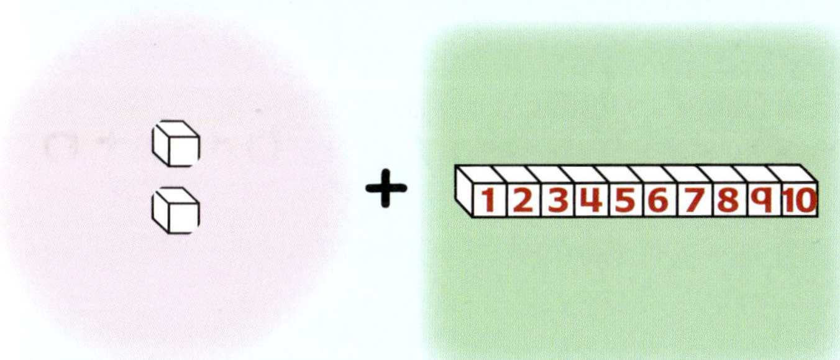

______ + ______ = ______

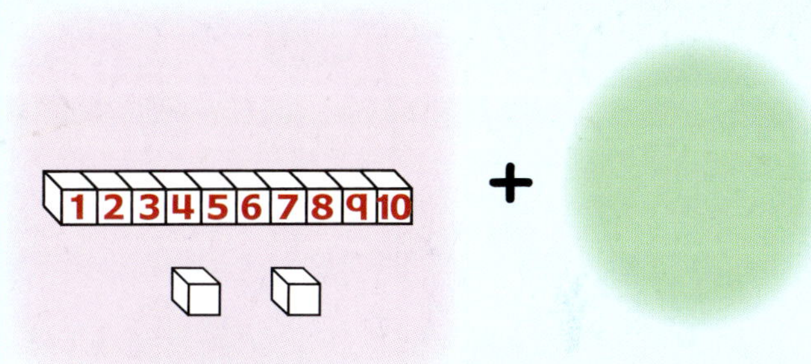

______ + ______ = ______

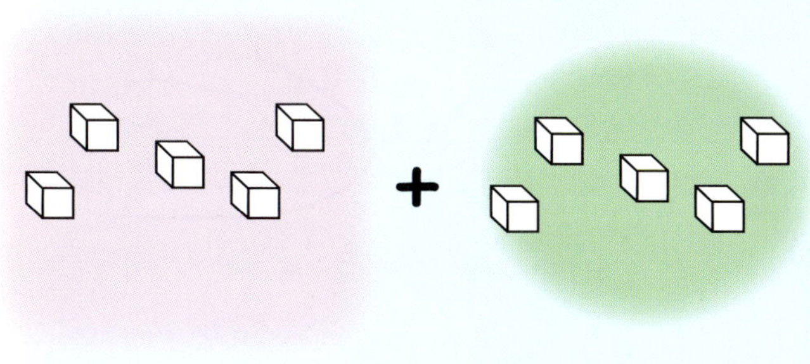

______ + ______ = ______

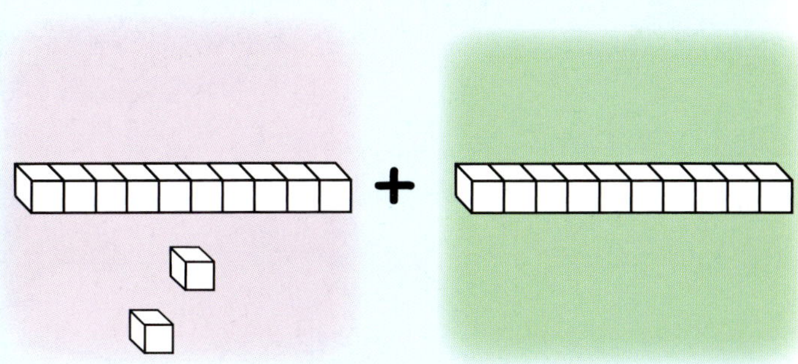

______ + ______ = ______

Do the actions and count aloud, repeating each pattern five times.

	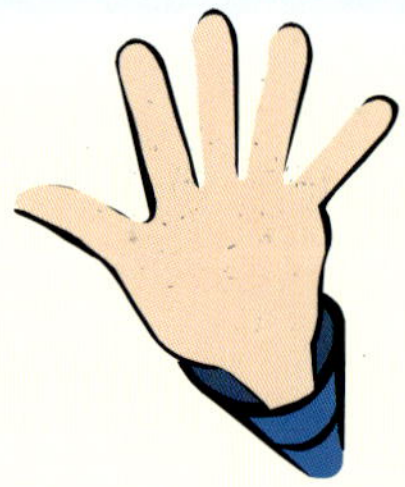		
TAP	WAVE	STRETCH	BEND

Put an X on all the objects with three or more corners that are triangles or red.

Put an X on all the objects that are yellow or circles.

Put an X on all the objects that are green or square.

Put an X on all the shapes that have corners.

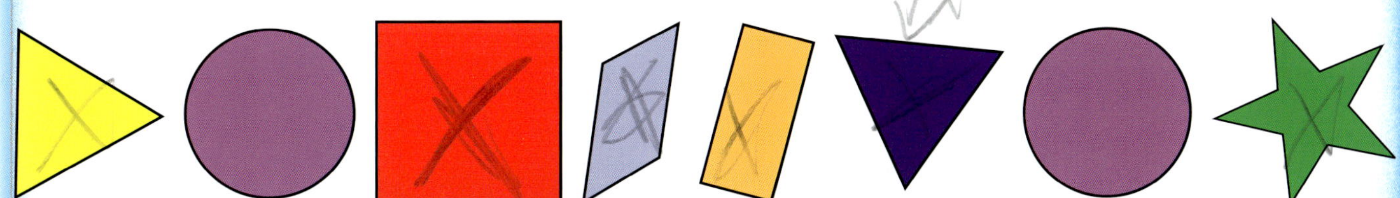

Put an X on all the rectangles.

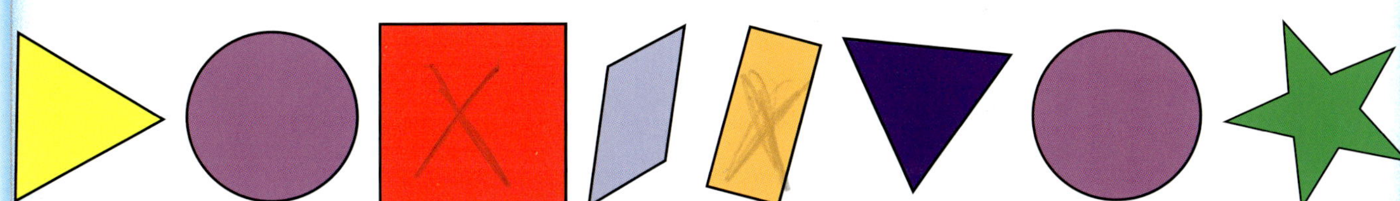

NOTE: Marking the square as a rectangle is acceptable since all squares are rectangles. Although, at this point, a child might not know all squares are rectangles.

How many of each toy is in the store?

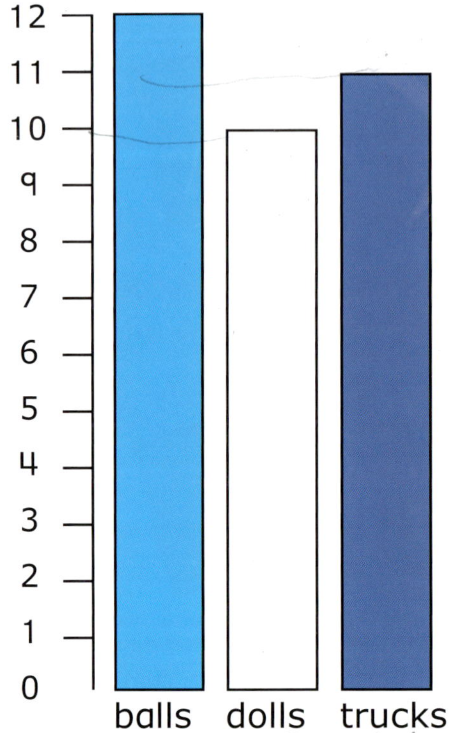

How many of each thing does this person have?

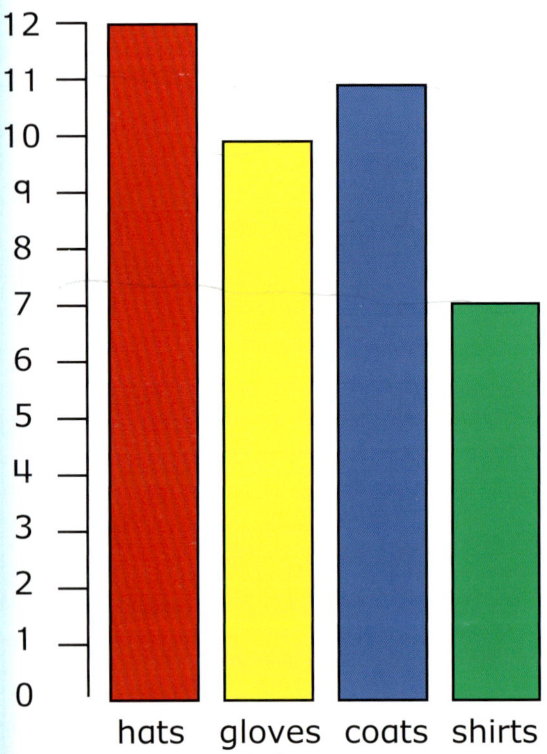

How many of each school supply does each student have?

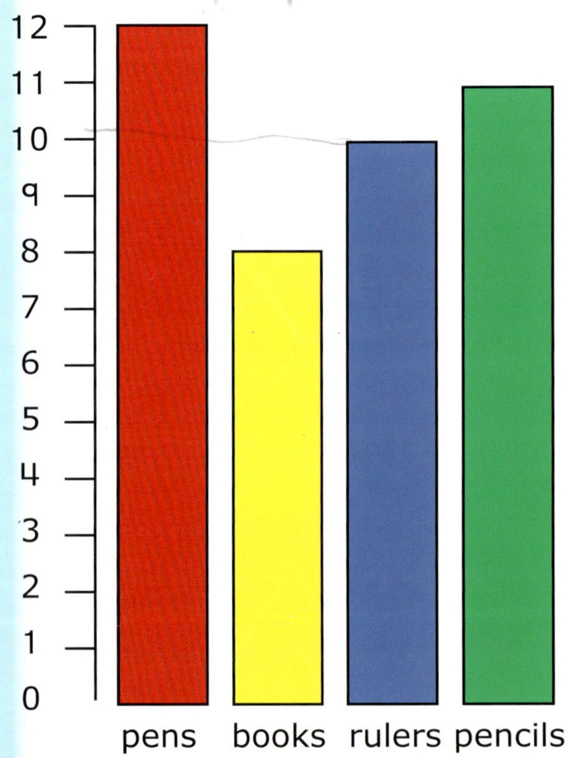

A store has 10 dogs and 12 cats.
Create a bar graph that shows this.

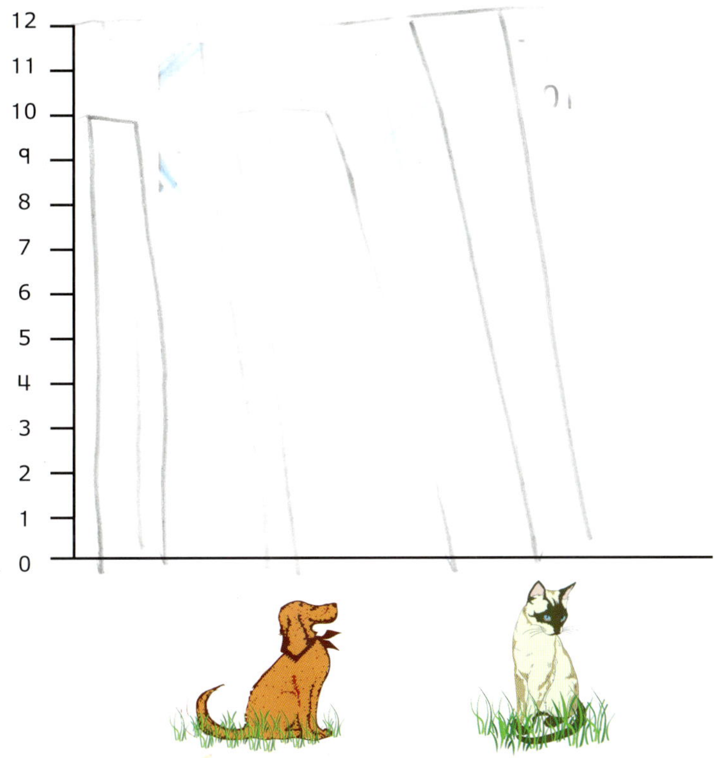

A store has 8 dogs and 2 cats.
Create a bar graph that shows this.

Which is longer, your arm or your foot?

Which is shorter, your bed or your leg?

Which is taller, a horse or a dog?

Which is shorter, a flagpole or the flag that goes on it?

Which is shorter, a baseball bat or a slipper?

Which is longer, your shoe or your arm?

Which is higher, the roof of a car or the seat on a bike?

Which takes longer, eating lunch or watching a movie?

Complete the tally marks to find the solution. Then write each number sentence.

|||||| + |||| = ( )

___ + ___ = ___

𝍸 + 𝍸 = ( )

___ + ___ = ___

||||||| + 𝍸 = ( )

___ + ___ = ___

|||||||| + |||| = ( )

___ + ___ = ___

|||||| + ||||| = ( )

___ + ___ = ___

Trace the numerals, then write the next three numerals that would continue the pattern.

6 5 11 6 5 11 6 5 11

7 5 12 7 5 12 7 5 12

5 5 10 5 5 10 5 5 10

6 6 12 6 6 12 6 6 12

10 2 12 10 2 12 10 2 12

4 8 12 4 8 12 4 8 12

Complete the tally marks to find the solution. Then write each number sentence.

𝍸 + |||||| =

____ ____ = ____

||||||||||| + =

____ ____ = ____

𝍸 | + ||||| =

____ ____ = ____

|| 𝍸 + || =

____ ____ = ____

|||||| + 𝍸 | =

____ ____ = ____

Put an X on all the triangles that are not blue.

Put an X on all the circles and blue objects.

Put an X on all the squares and purple objects.

Put an X on all the shapes that have corners.

Put an X on all the rectangles.

NOTE: Marking the square as a rectangle is acceptable since all squares are rectangles. Although, at this point, a child might not know all squares are rectangles.

A ball team has 11 gloves, 12 balls, and 10 bats. Create a bar graph that shows this.

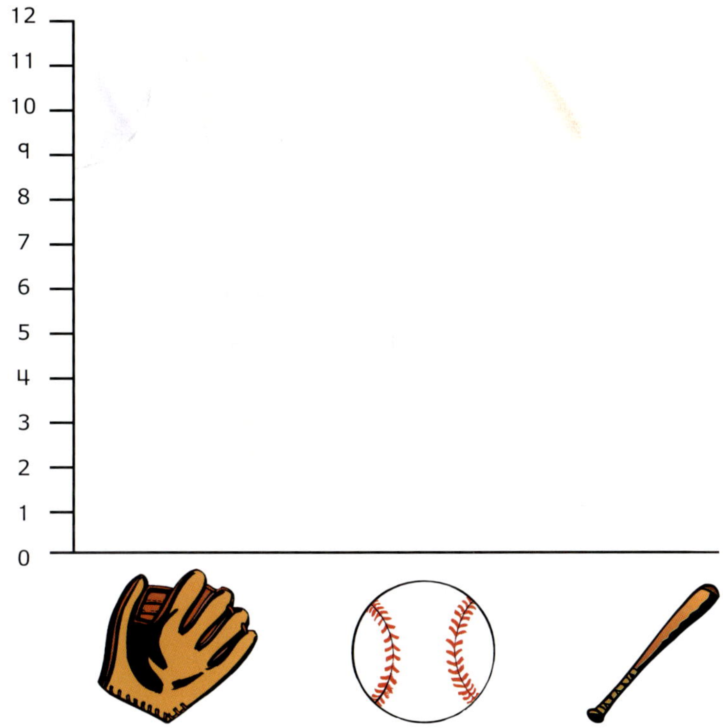

A shop has 9 bikes, 12 helmets, 11 pairs of shoes, and 10 skateboards. Create a bar graph that shows this.

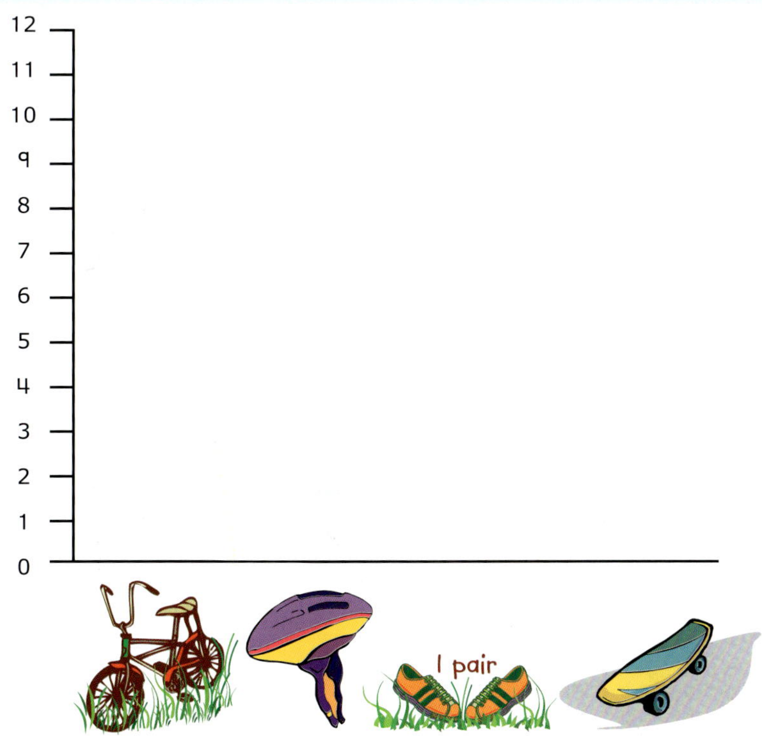

The number 33 comes after 32.

32 33

Trace each numeral, then write the numeral that comes after it.

43 ______ 50 ______

64 ______ 72 ______

58 ______ 35 ______

81 ______ 76 ______

56 ______ 40 ______

39 ______ 62 ______

92 ______ 98 ______

Use the number lines below to solve each problem.

0 1 2 3 4 5 6 7 8 9 10 11 12

3 + 8 = ______

0 1 2 3 4 5 6 7 8 9 10 11 12

5 + 6 = ______

0 1 2 3 4 5 6 7 8 9 10 11 12

7 + 4 = ______

0 1 2 3 4 5 6 7 8 9 10 11 12

3 + 7 = ______

0 1 2 3 4 5 6 7 8 9 10 11 12

2 + 10 = ______

0 1 2 3 4 5 6 7 8 9 10 11 12

3 + 9 = ______

Trace the numerals, then write the next three numerals that would extend the pattern.

8 3 11 8 3 11 8 3 11 ___ ___ ___

7 3 10 7 3 10 7 3 10 ___ ___ ___

3 9 12 3 9 12 3 9 12 ___ ___ ___

8 4 12 8 4 12 8 4 12 ___ ___ ___

4 8 12 4 8 12 4 8 12 ___ ___ ___

10 0 10 0 10 0 10 0 10 ___ ___ ___

Use the number lines below to solve the problems.

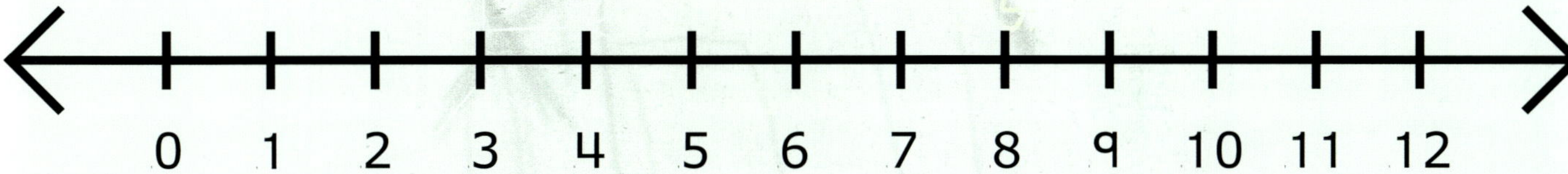

7 + 5 = ______

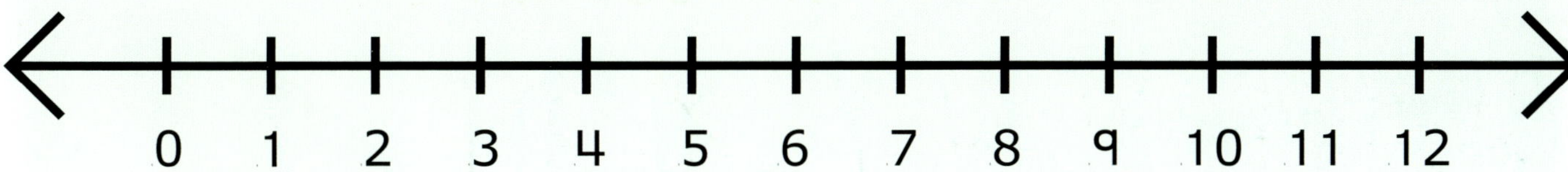

5 + 7 = ______

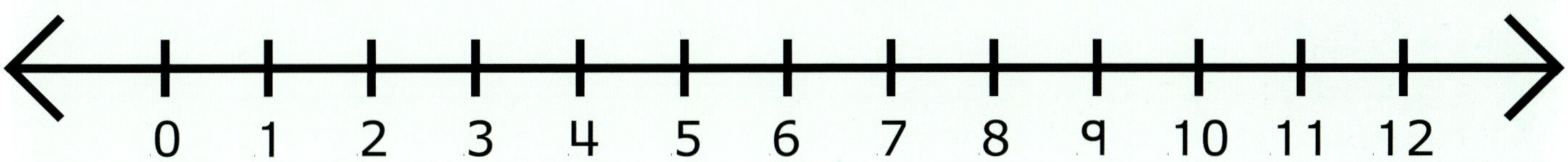

6 + 6 = ______

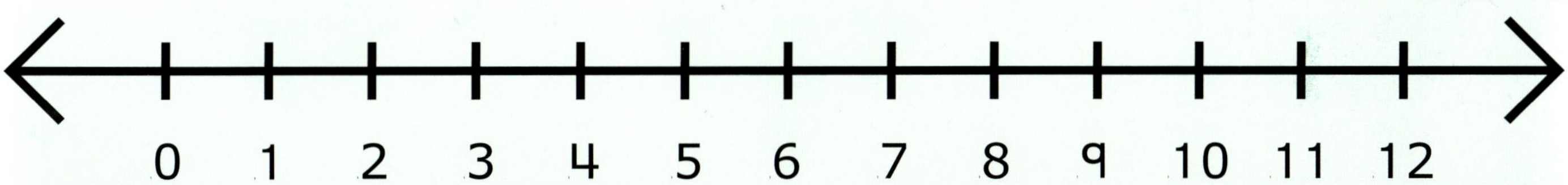

11 + 1 = ______

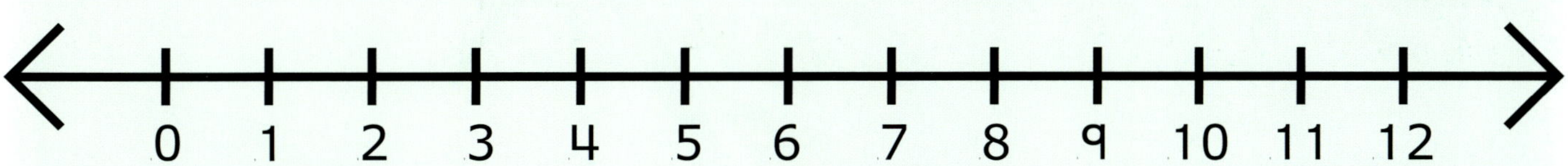

1 + 11 = ______

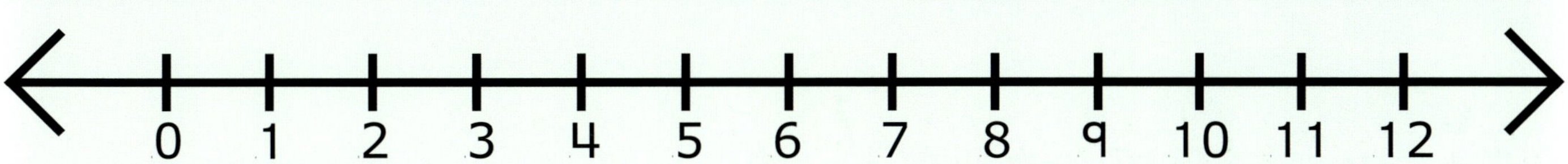

9 + 2 = ______

Cross out the shape that doesn't belong.

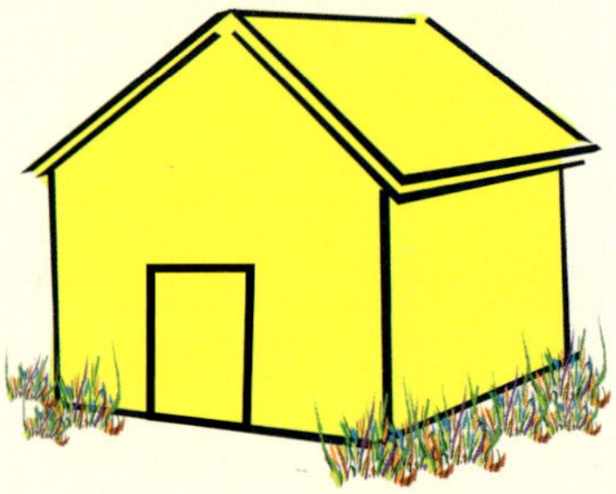
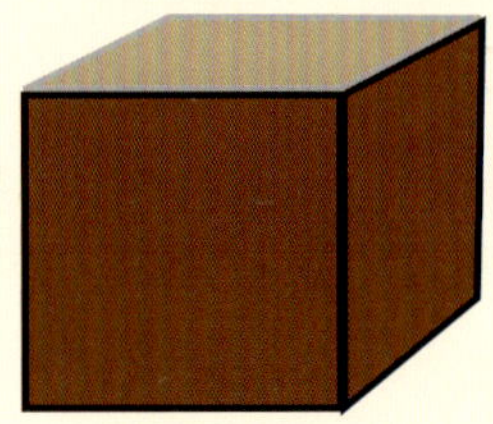

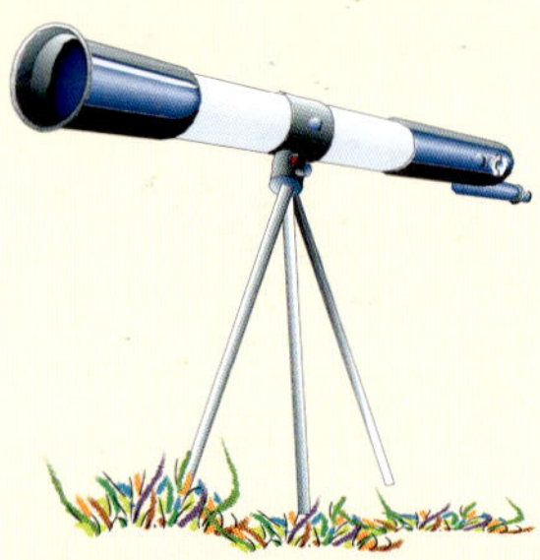

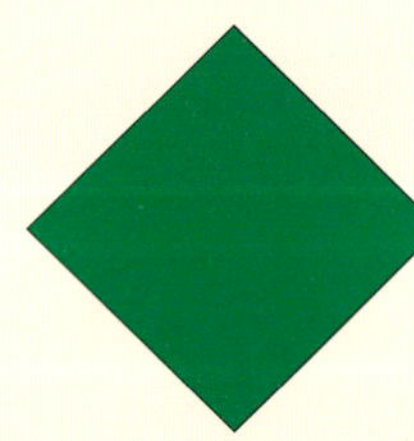

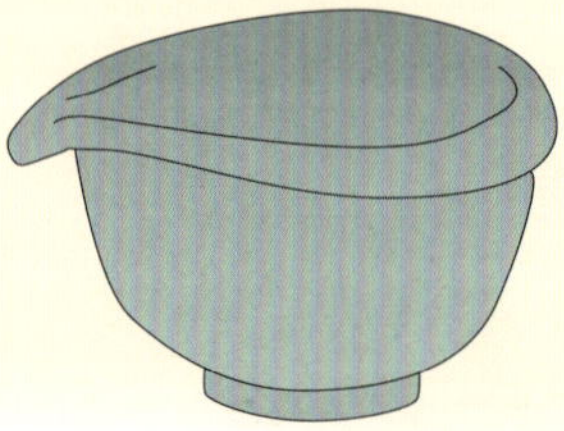
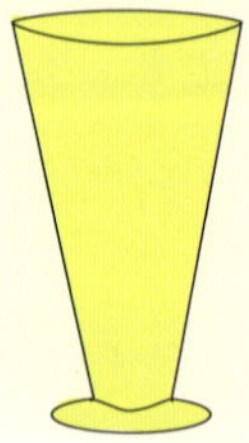
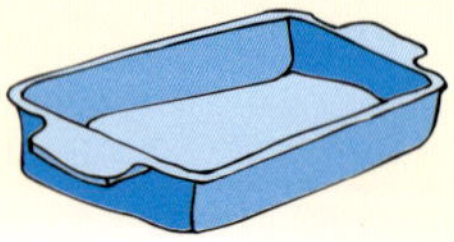

Write each number sentence.

You have 5 toys and get 7 more. How many toys do you have after that?

________ + ________ = ________

There are 7 books on one shelf and five more books on another shelf. How many books are there in all?

________ + ________ = ________

You have your bike. Nine friends bring their bikes to your house, too. How many bikes are at your house?

________ + ________ = ________

You have 2 pennies. You find 3 pennies. Later you find a nickel. How much money do you have?

_____ + _____ + _____ = ________

The number 5 comes before 6.

5 6

Trace each numeral, then write the numeral that comes before it.

____ 4	____ 2	____ 5
____ 8	____ 7	____ 9
____ 10	____ 11	____ 12
____ 3	____ 14	____ 13
____ 15	____ 17	____ 16
____ 19	____ 18	____ 20
____ 25	____ 21	____ 23
____ 24	____ 22	____ 26

Add the number of bulbs in each set of seasonal lights. Then write the number sentence for each problem below.

_____ + _____ = _____

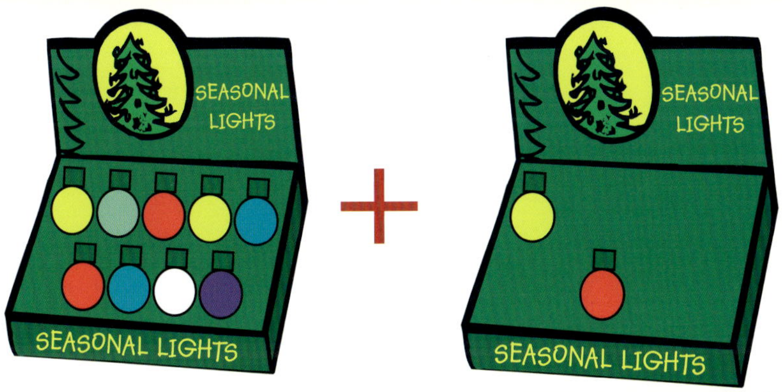

_____ + _____ = _____

_____ + _____ = _____

_____ + _____ = _____

Trace the numerals and do the problems. Go in order from left to right. Then, on the next page, connect the answer dots to find the mystery object.

4 + 7 = ______ 5 + 5 = ______

5 + 6 = ______ 6 + 6 = ______

8 + 3 = ______ 7 + 3 = ______

0 + 8 = ______ 5 + 7 = ______

9 + 2 = ______ 8 + 4 = ______

5 + 7 = ______ 4 + 6 = ______

6 + 2 = ______ 2 + 8 = ______

6 + 5 = ______ 12 + 0 = ______

8 + 3 = ______ 10 + 1 = ______

10 + 2 = ______ 10 + 0 = ______

My top is white and fluffy and I come in many flavors. You can eat me with a straw or a spoon.

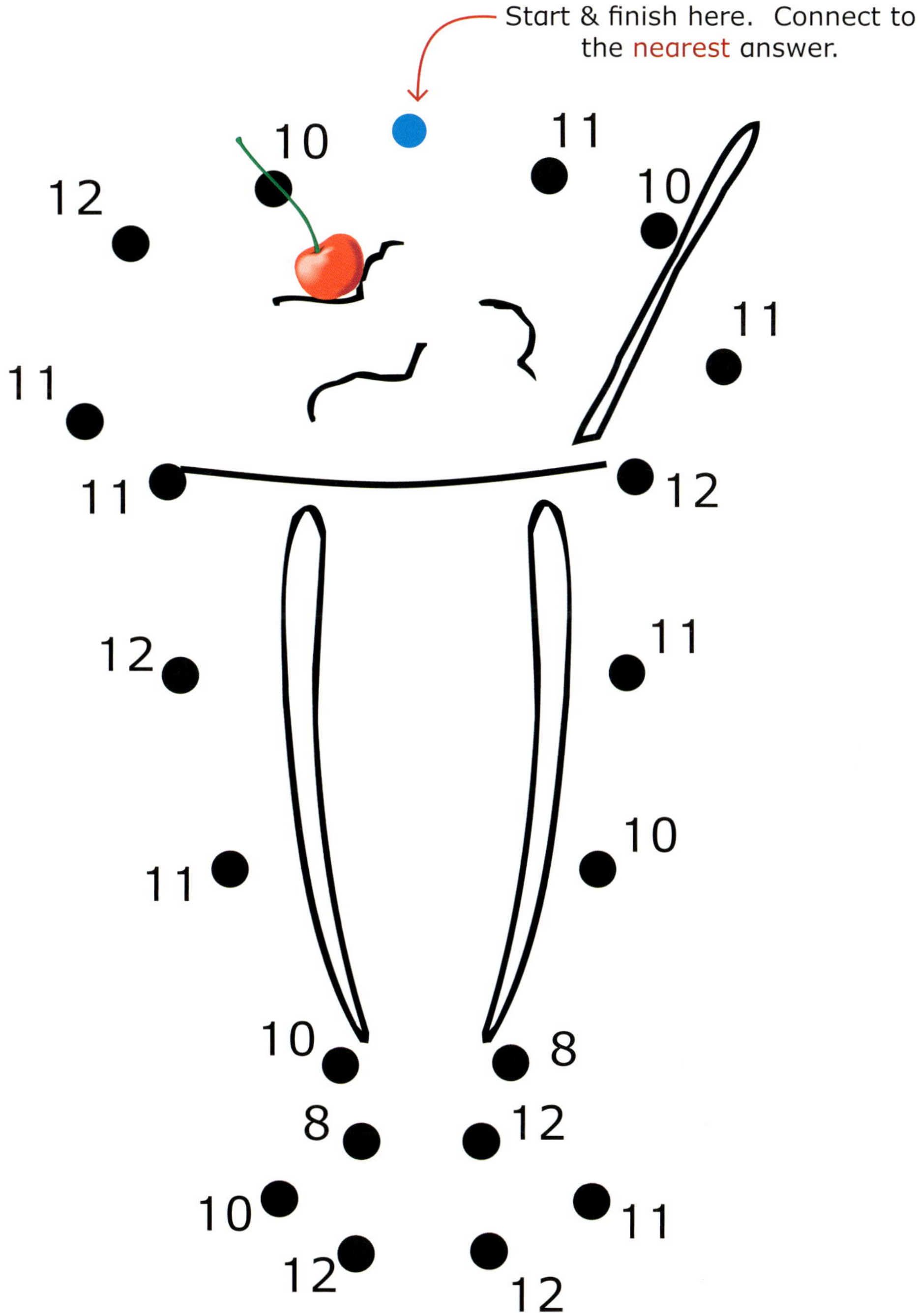

Color the mystery object and add something new to the picture.

Write the number sentence for each problem below.

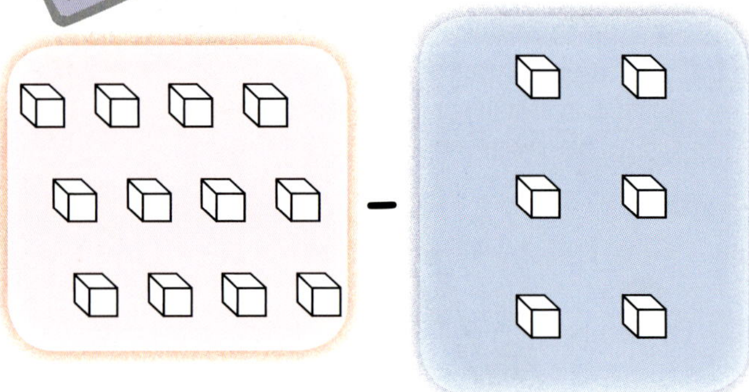

______ - ______ = ______

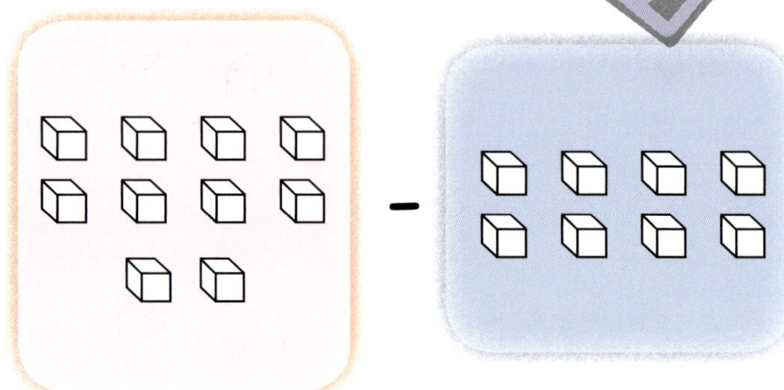

______ - ______ = ______

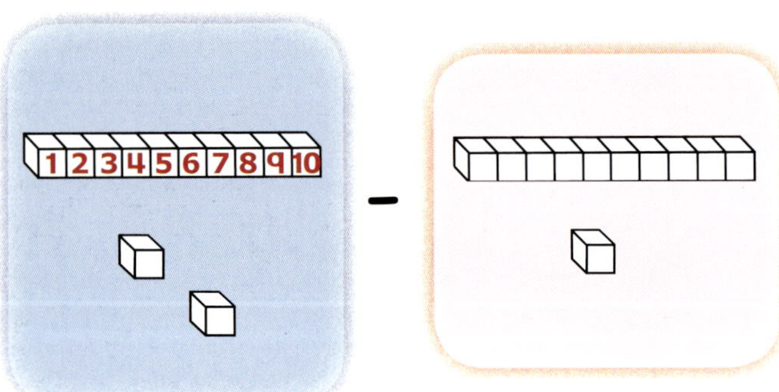

______ - ______ = ______

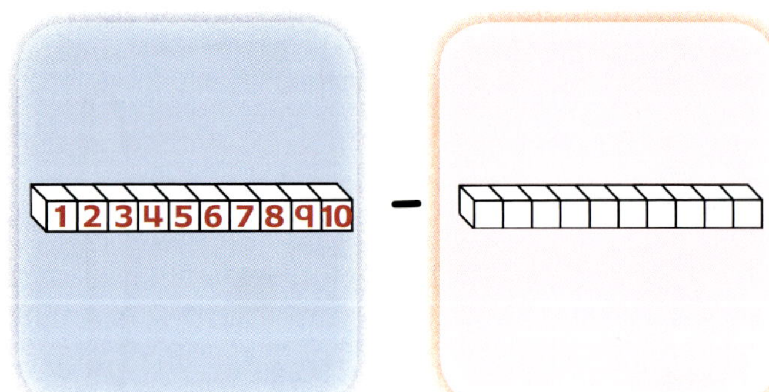

______ - ______ = ______

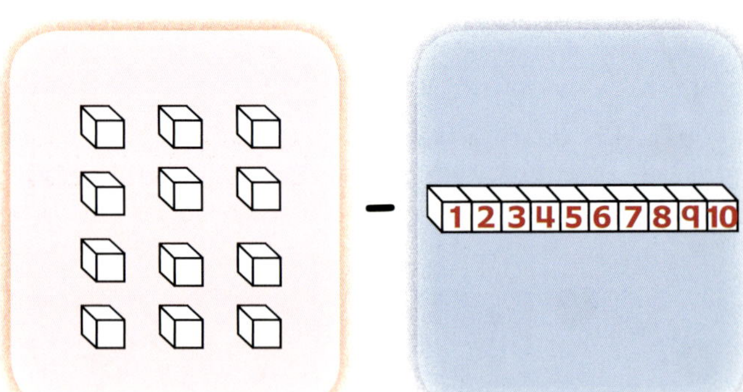

______ - ______ = ______

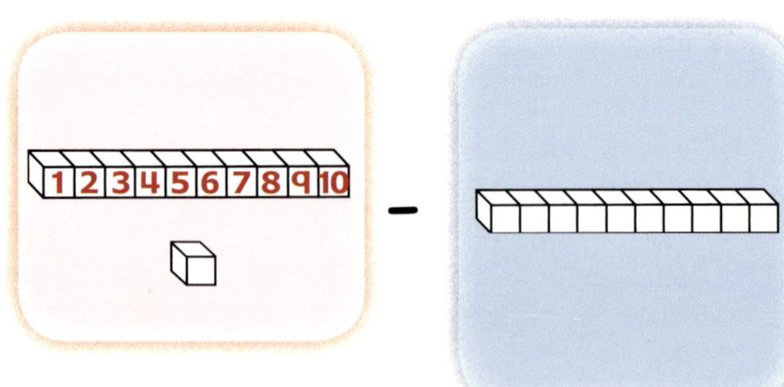

______ - ______ = ______

Write the number sentence for each problem below. Be sure to include a minus sign.

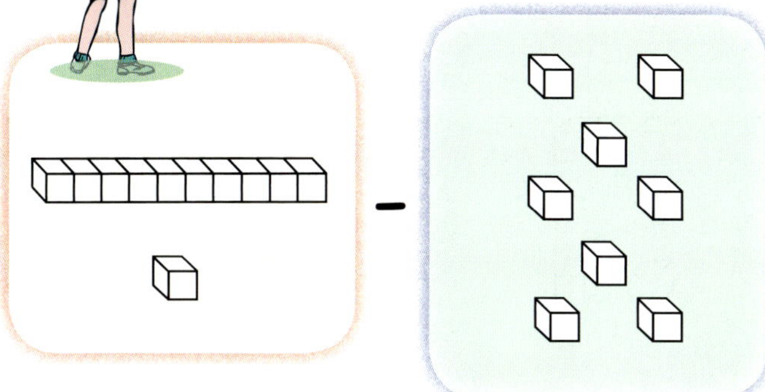

______ ______ = ______

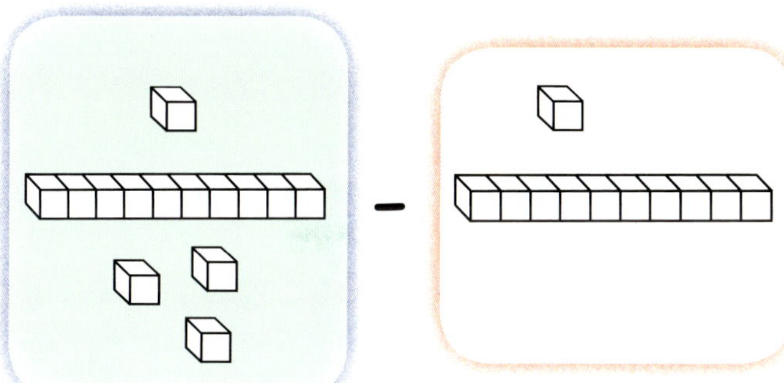

______ ______ = ______

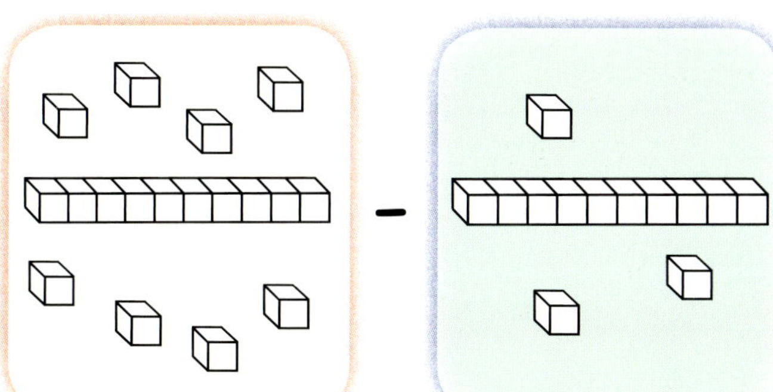

______ ______ = ______

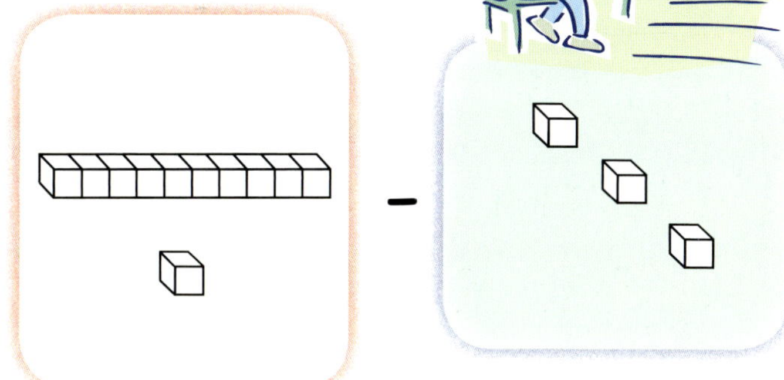

______ ______ = ______

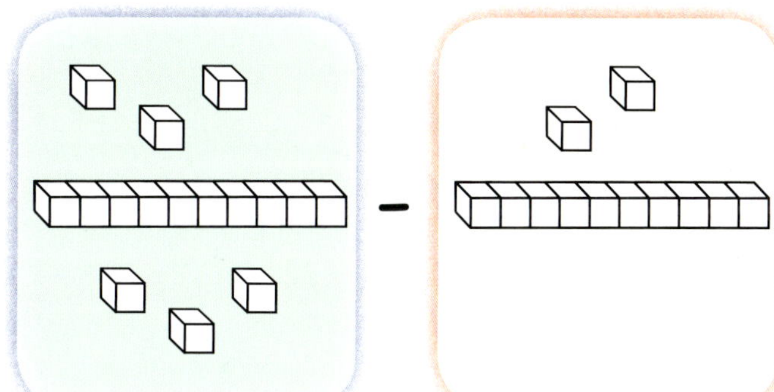

______ ______ = ______

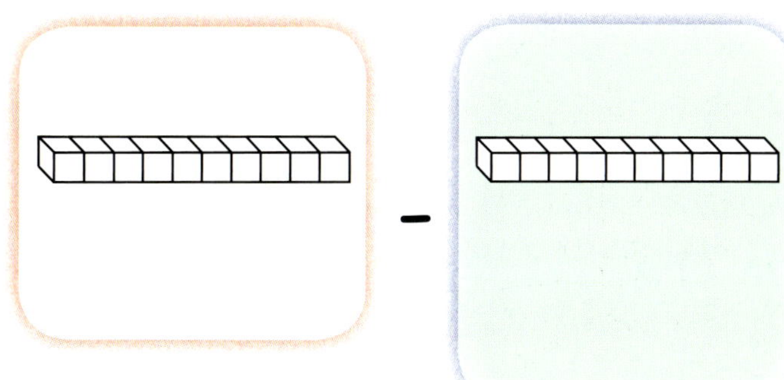

______ ______ = ______

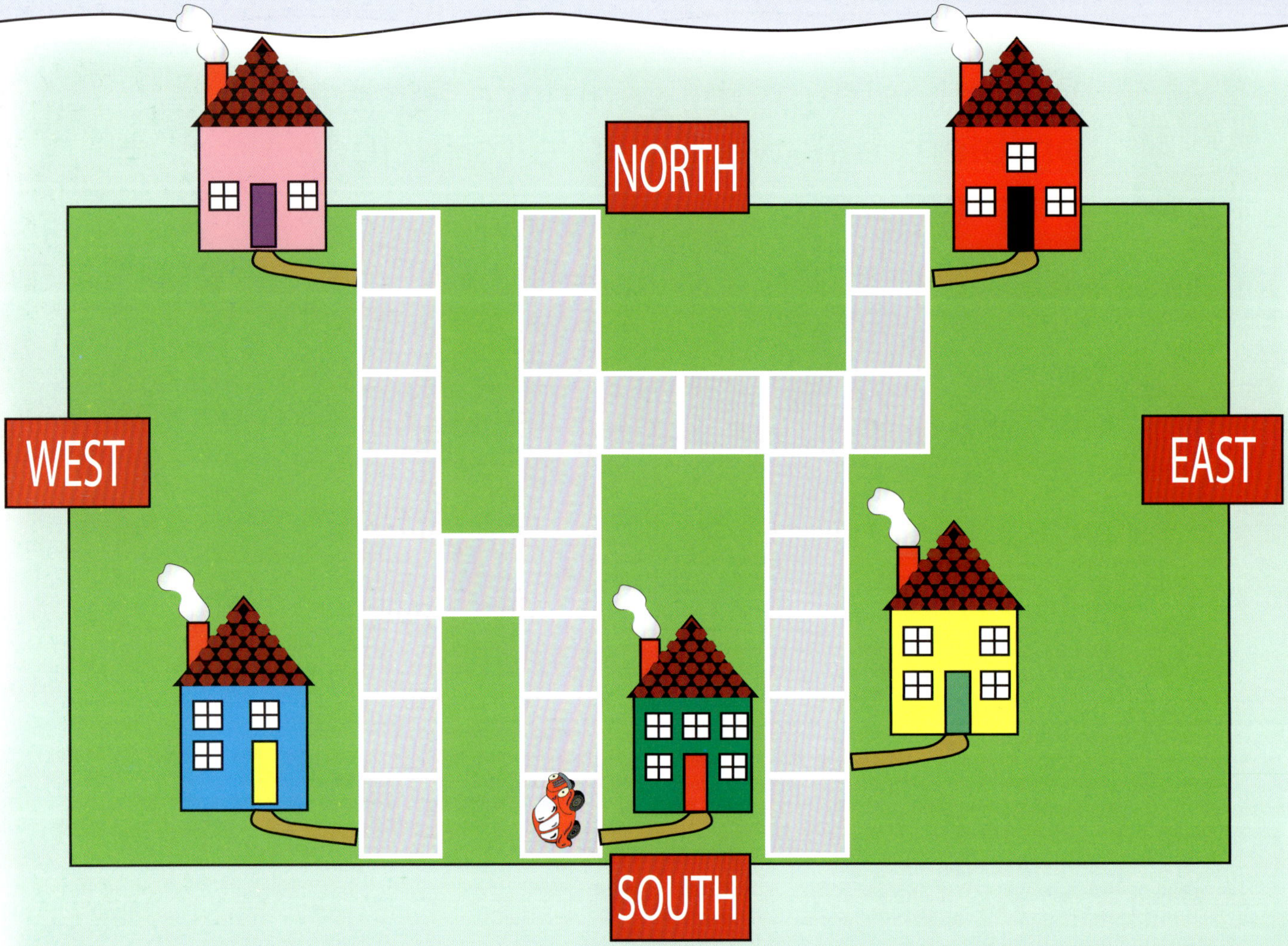

Give the red car road directions from the green house to the nearest house by road.

Go ______________ ______________ squares. Then go
(compass direction) (# of squares)

______________ ______________ squares. Then go
(compass direction) (# of squares)

______________ ______________ squares. You are there.
(compass direction) (# of squares)

Give the red car road directions from the green house to the farthest house by road.

Go ______________ ______________ squares. Then go
(compass direction) (# of squares)

______________ ______________ squares. Then go
(compass direction) (# of squares)

______________ ______________ squares. You are there.
(compass direction) (# of squares)

Complete the tally marks to find the solution. Then write each number sentence. Be sure to include a minus sign.

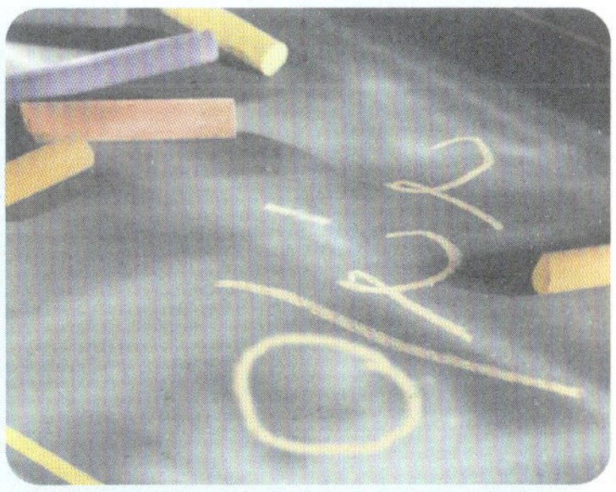

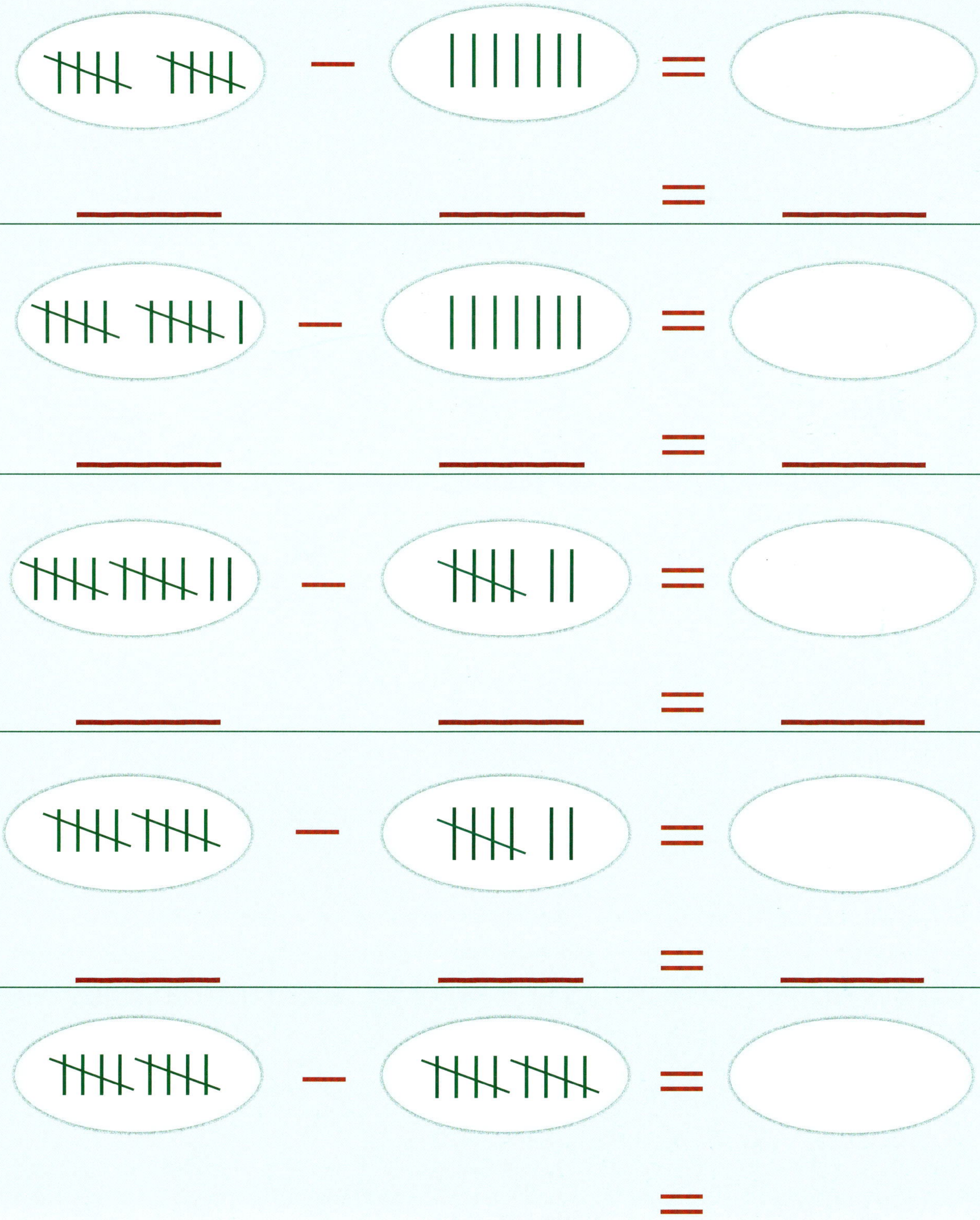

Write the number sentence for each of the following problems. Be sure to include a minus sign.

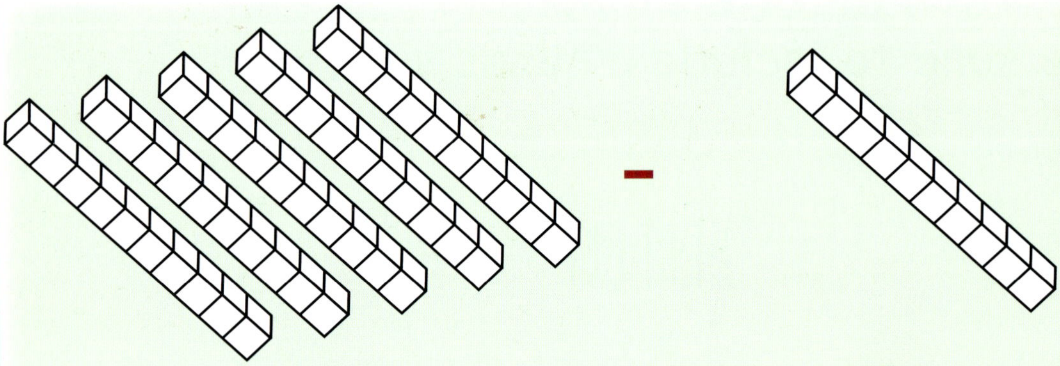

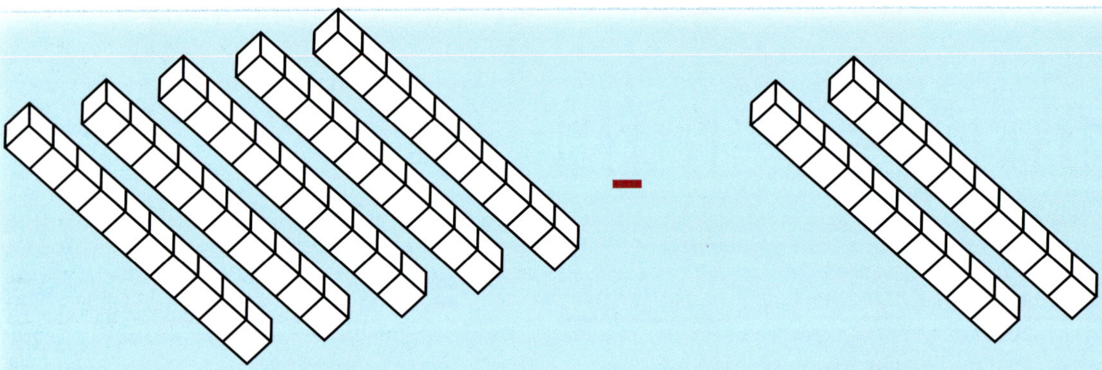

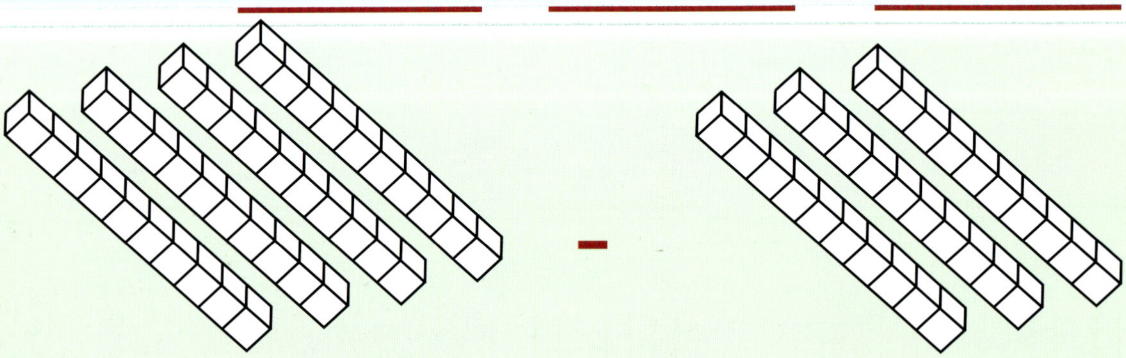

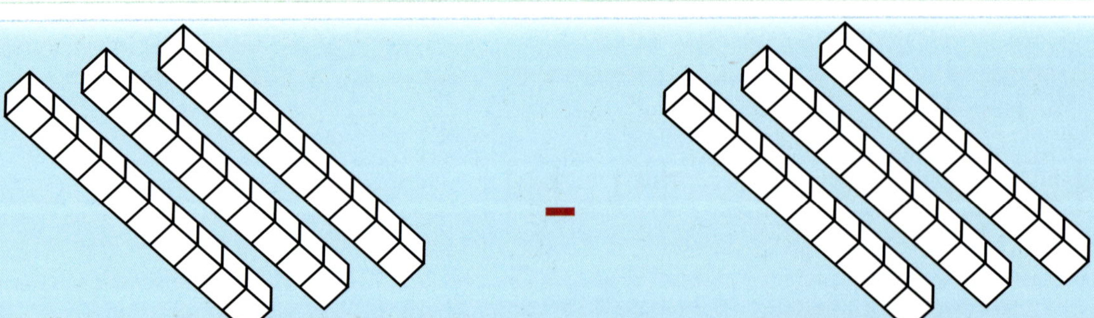

Write the number sentence for each problem.

If you give away one pumpkin, how many will be left on this table?

______ ______ = ______

If you give away eight pumpkins, how many will be left on this table?

______ ______ = ______

If you give away two pumpkins, how many will be left on this table?

______ ______ = ______

If you give away four pumpkins, how many will be left on this table?

______ ______ = ______

Complete the tally marks to find the solution to each number sentence. Then write the number sentence.

− =

____ ____ = ____

− =

____ ____ = ____

− =

____ ____ = ____

− =

____ ____ = ____

− =

____ ____ = ____

Use the number line below to solve the problems.

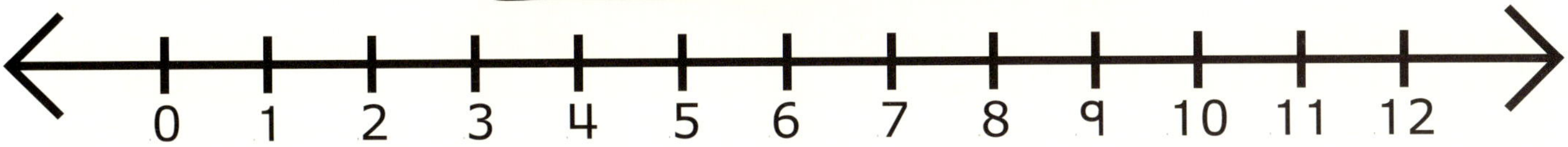

10 - 3 = ______

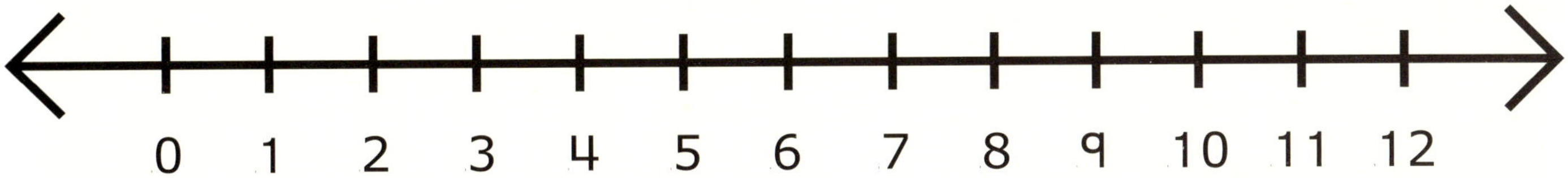

11 - 5 = ______

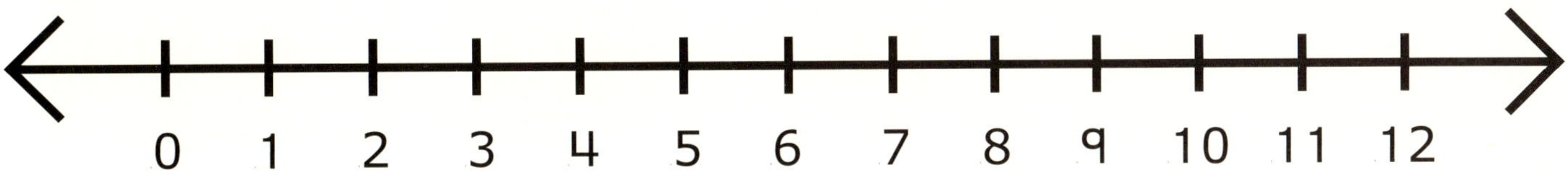

11 - 6 = ______

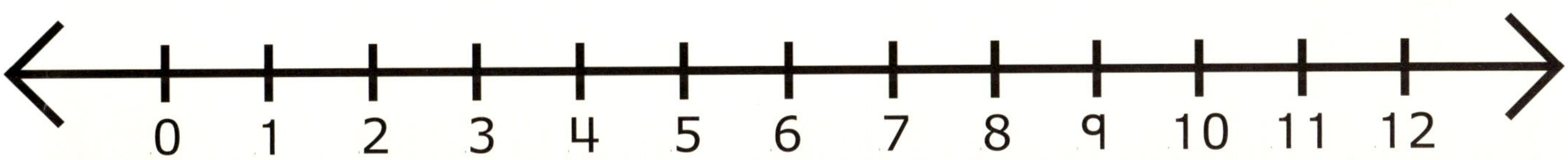

12 - 2 = ______

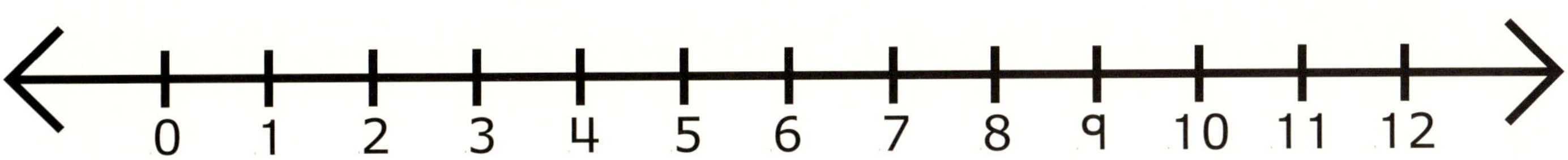

12-8 = ______

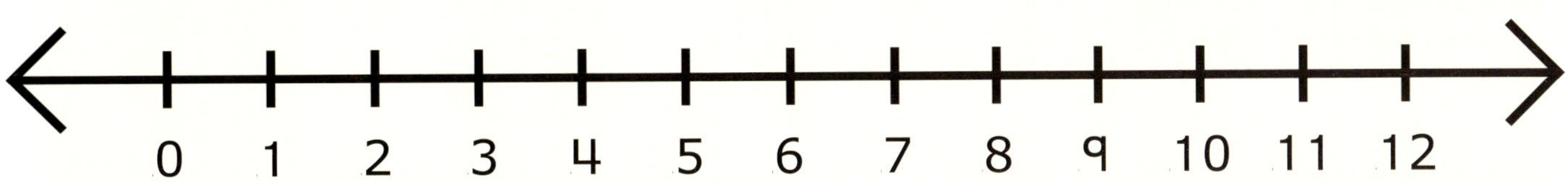

10 - 7 = ______

Circle the shape that the object in each blue circle will make when it is folded.

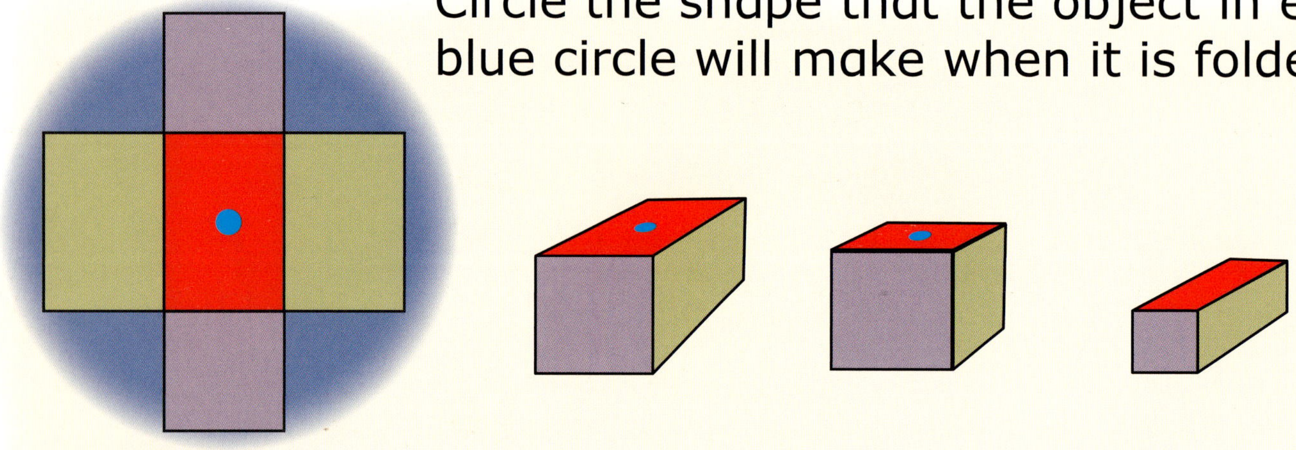

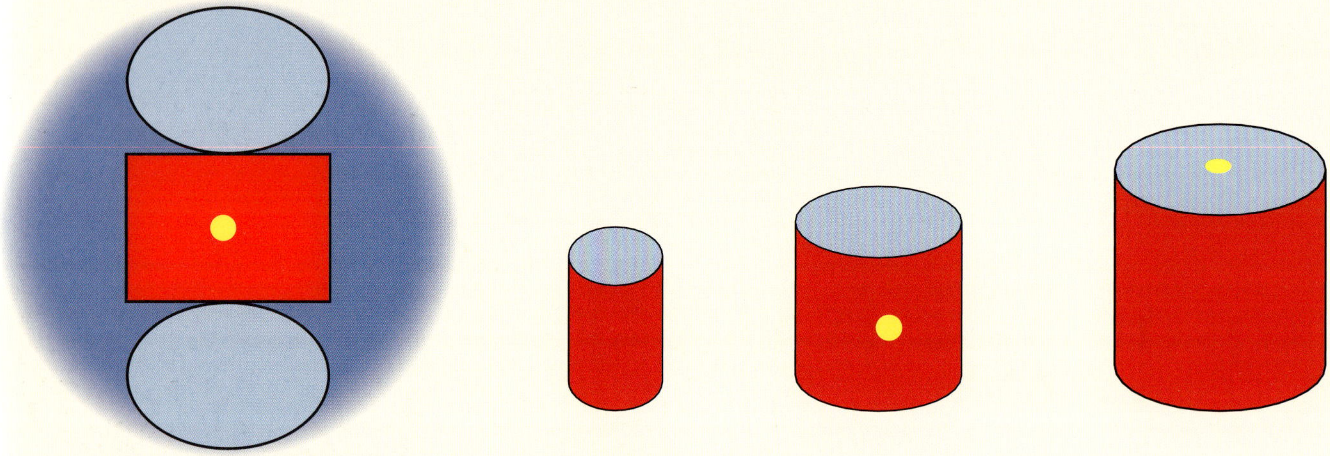

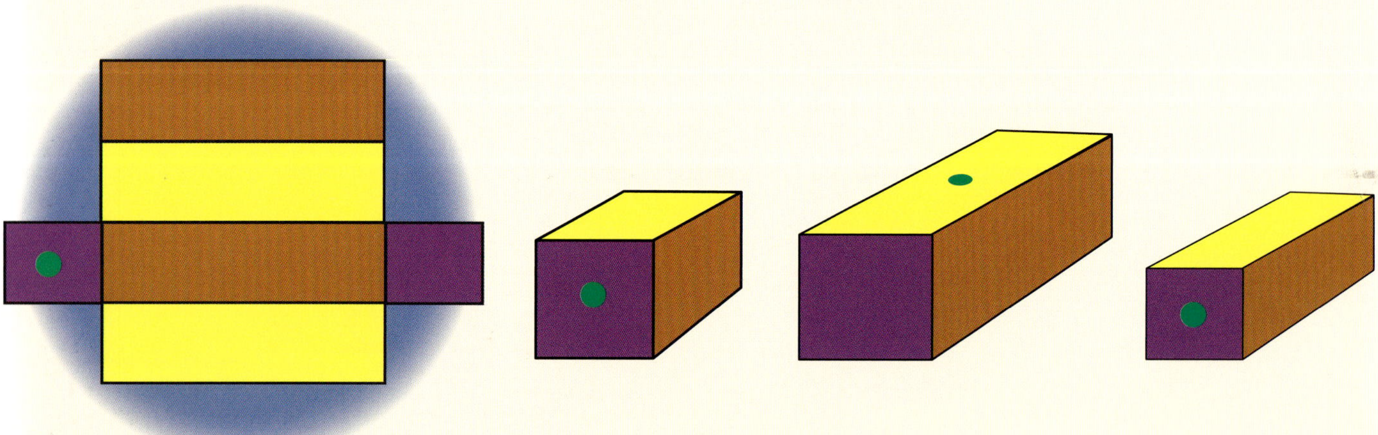

Use the number lines below to solve the problems.

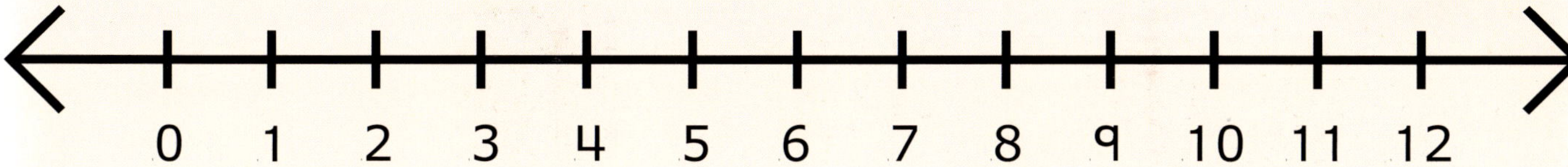

10 - 3 = ______

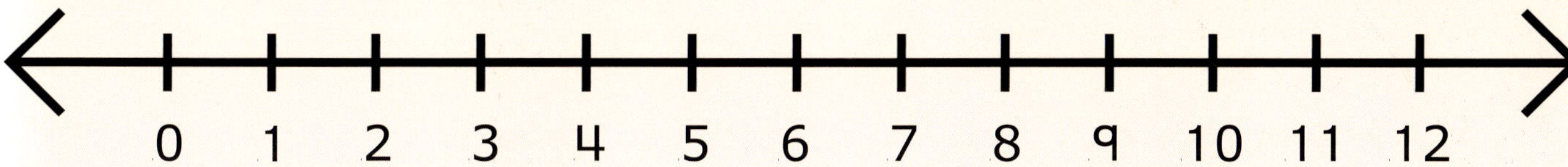

11 - 7 = ______

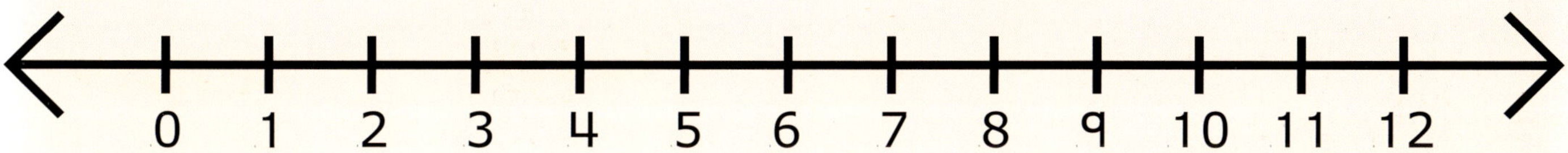

12 - 1 = ______

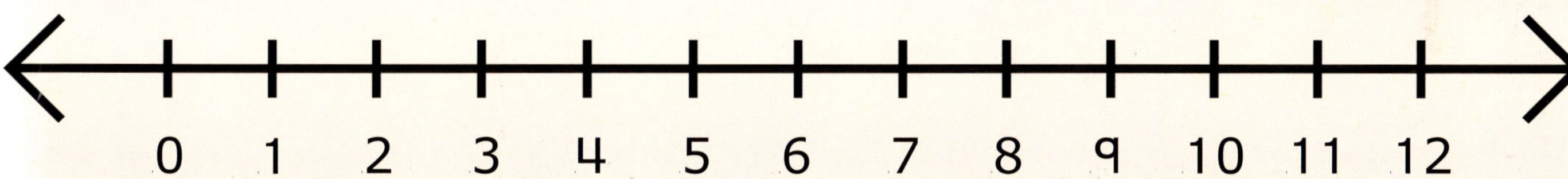

11 - 9 = ______

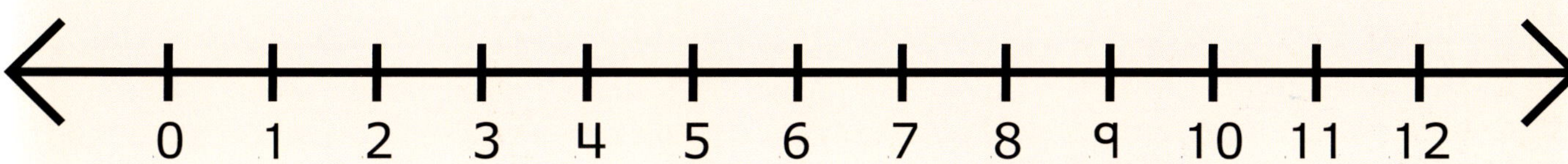

10 - 6 = ______

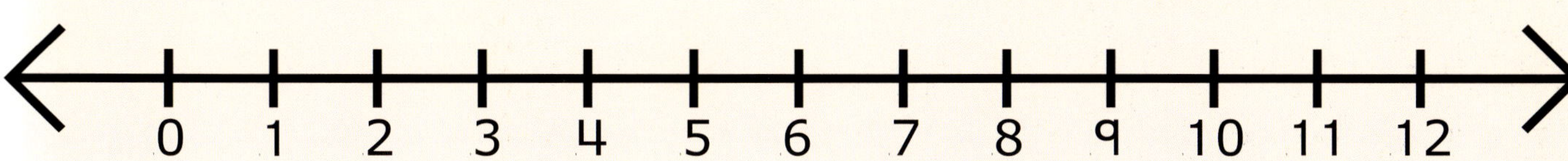

12 - 6 = ______

Parts of a shape or object split into three equal parts are called thirds. Each of the 3 equal parts is called one third.

Circle objects split into thirds.

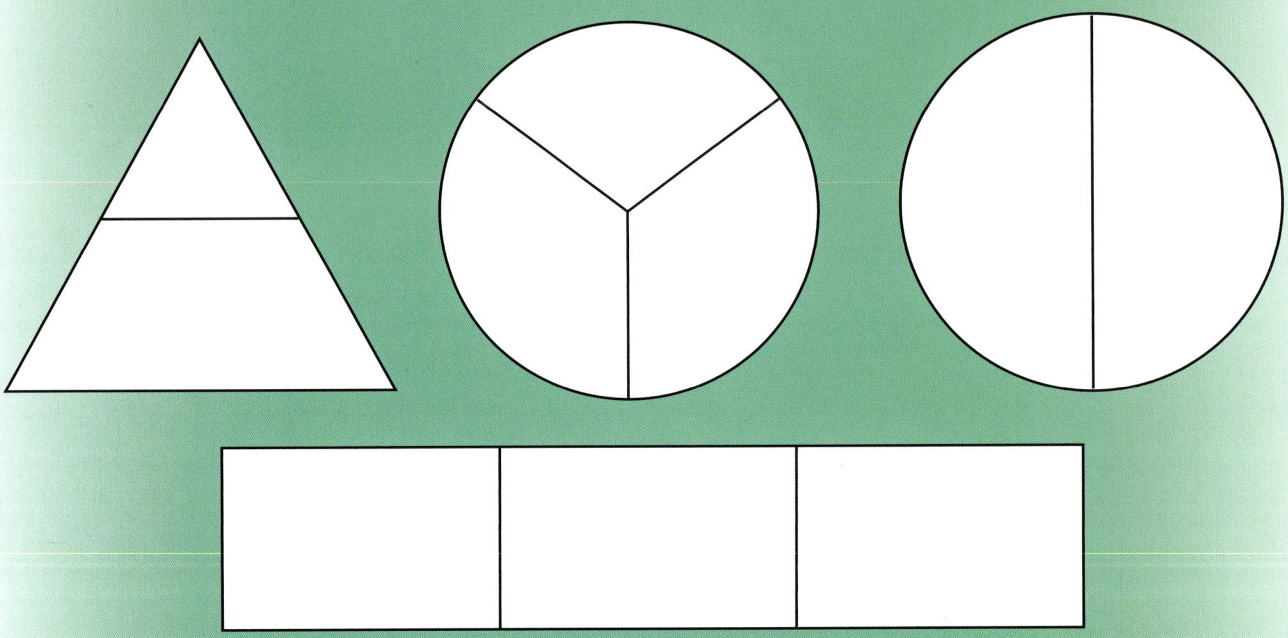

Color $\frac{1}{3}$ of each.

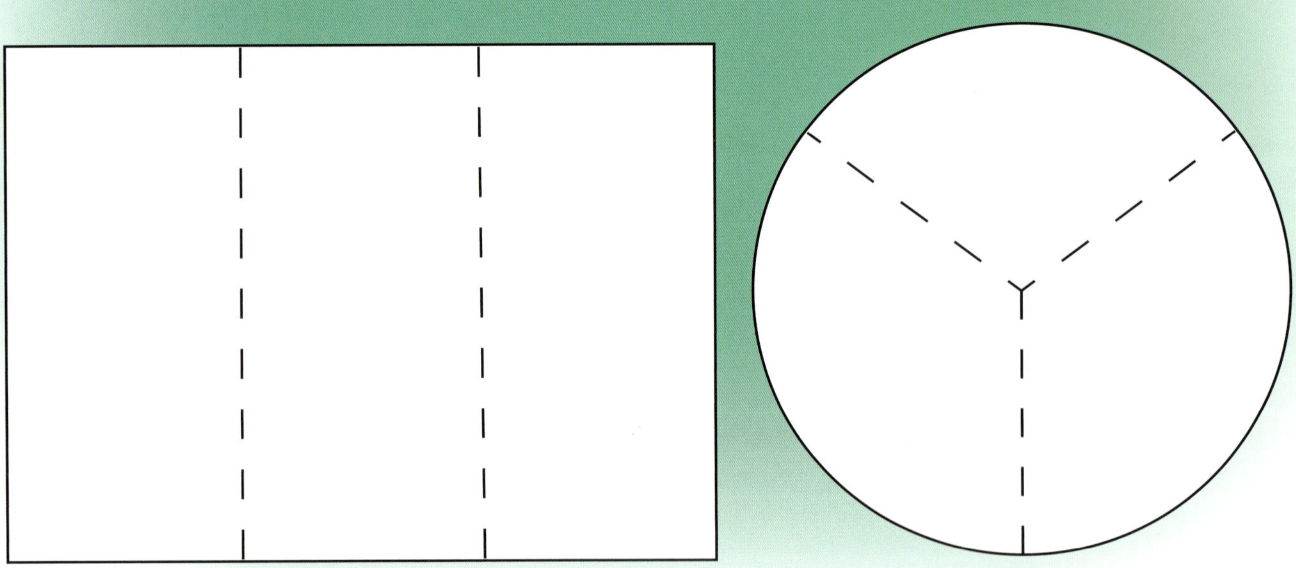

The biggest box will hold 30 balls when full. The smallest box will hold 10 balls when full.

1. Which two boxes will hold the most balls? ______________
2. Which two boxes will hold the fewest balls? ____________
3. What is the most balls the green box could hold? _________

   Explain your thinking.
4. Point to all the boxes that could hold 15 balls.
5. Point to all the boxes that could hold 9 balls.
6. Point to all the boxes that could hold 19 balls.

Do the problems. Go in order from left to right. Then, on the next page, connect the answer dots to find the mystery animal.

12 - 2 = _____	11 - 2 = _____
10 - 2 = _____	12 - 8 = _____
12 - 7 = _____	10 - 0 = _____
10 - 3 = _____	11 - 6 = _____
11 - 2 = _____	12 - 5 = _____
12 - 1 = _____	11 - 5 = _____
10 - 1 = _____	11 - 3 = _____
10 - 5 = _____	10 - 6 = _____
10 - 7 = _____	10 - 8 = _____
10 - 9 = _____	12 - 0 = _____

I am the largest warm blooded mammal in the ocean. I eat by filtering silt at the bottom of the ocean through my teeth to find small plantlife for my food.

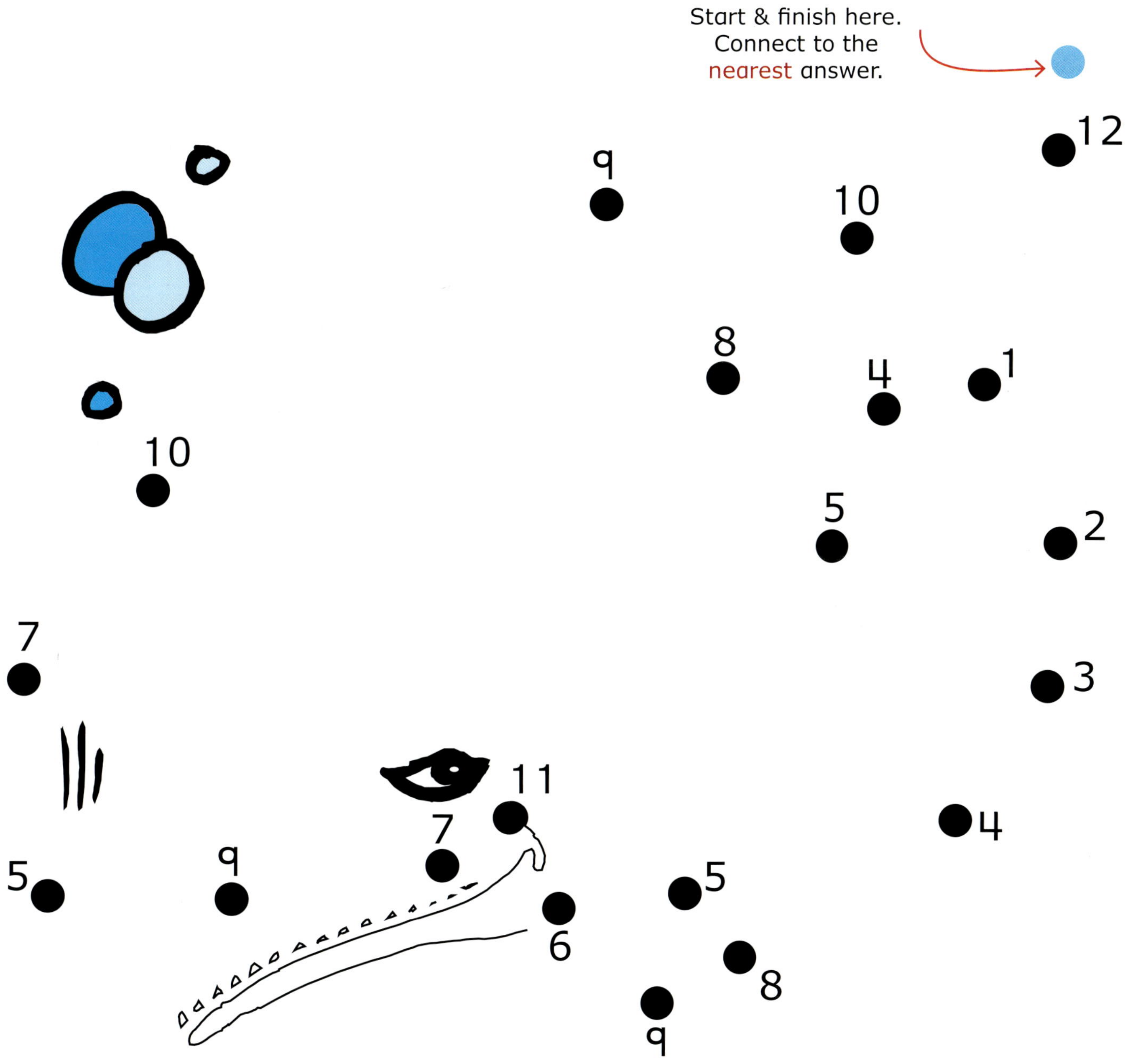

Color the mystery animal and add something new to the picture.

Write each number sentence.

You have 11 toys and give 3 to a friend. How many toys do you have after that?

________ - ________ = ________

There are 12 books on a shelf in your room. You return 3 to the library. How many books are left on the shelf ?

________ - ________ = ________

You have 11 party favors for your birthday party. You give out 4 party favors to your guests. How many party favors are left?

________ - ________ = ________

You have 12 pennies. You spend a nickel's worth of pennies. How many pennies do you have left?

________ - ________ = ________

Circle the coins you would use to make the amount in the blue circle.

23¢ =

15¢ =

18¢ =

20¢ =

Which is heavier, your arm or your foot?

Which is lighter, you or a baseball?

Who is heavier, your mom or you?

Which is lighter, a horse or a dog?

Which is lighter, 1 baseball bat or 4 baseball bats?

Which is heavier, all of your toys or 3 of your toys?

Which is heavier, a car or a bike?

Which is heavier, a school bus or a baseball?

Circle the season when leaves on trees turn colors.

Winter

Spring

Summer

Fall

Circle the season when leaves on trees start to grow.

Winter

Spring

Summer

Fall

Circle the season when there are no leaves on trees.

Winter

Spring

Summer

Fall

Circle the season when the leaves on the trees shade the playground.

Winter

Spring

Summer

Fall

Use the base 10 blocks to solve each problem. Then write and say each number sentence. Be sure to include a minus sign.

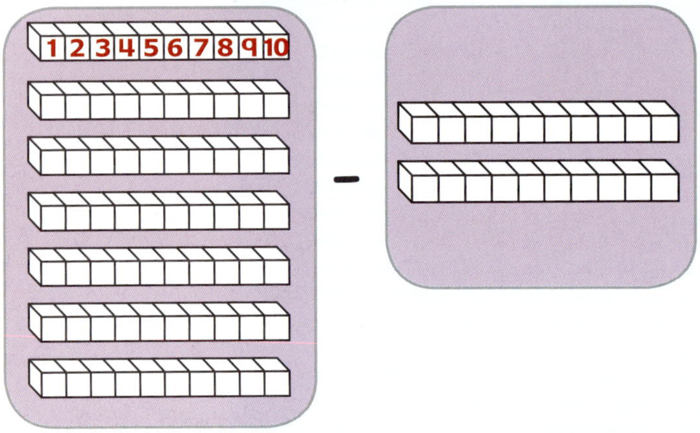

_____ _____ = _____

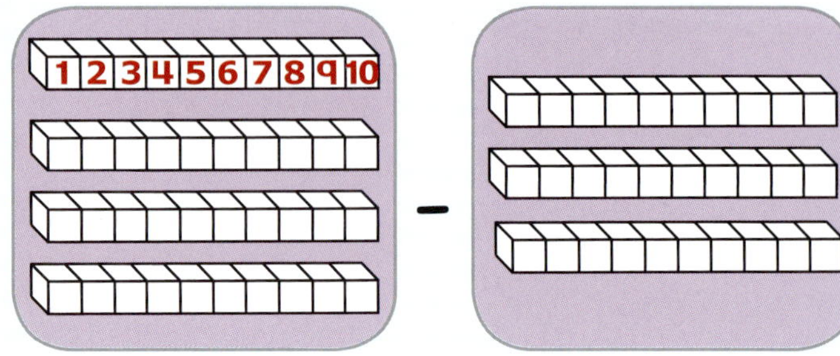

_____ _____ = _____

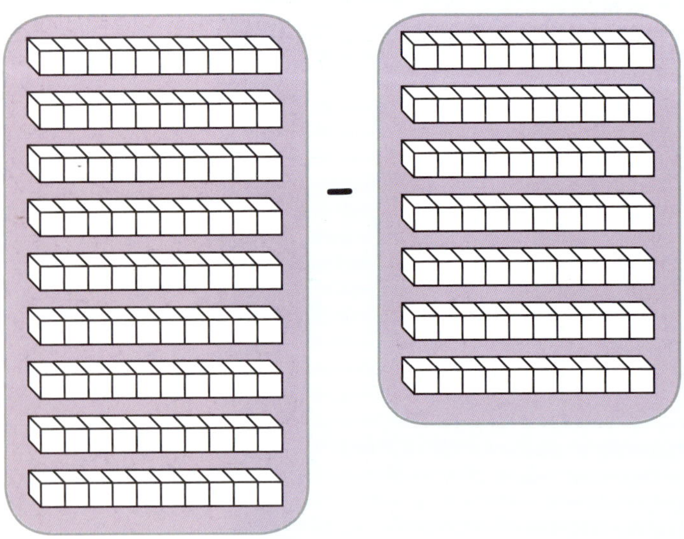

_____ _____ = _____

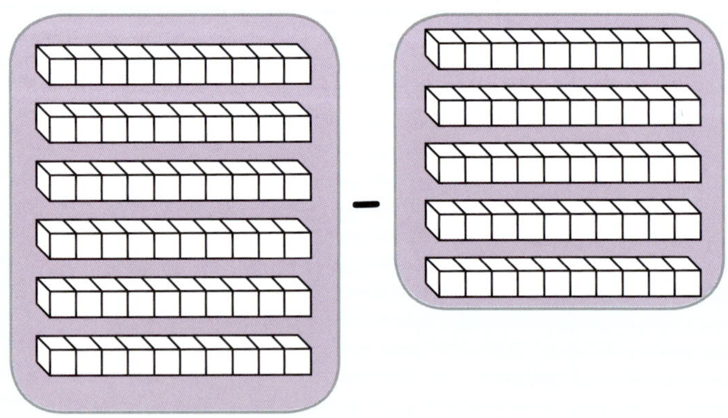

_____ _____ = _____

Use a toothpick to measure the length of the green line segment.

The green line segment is ________ toothpicks long.

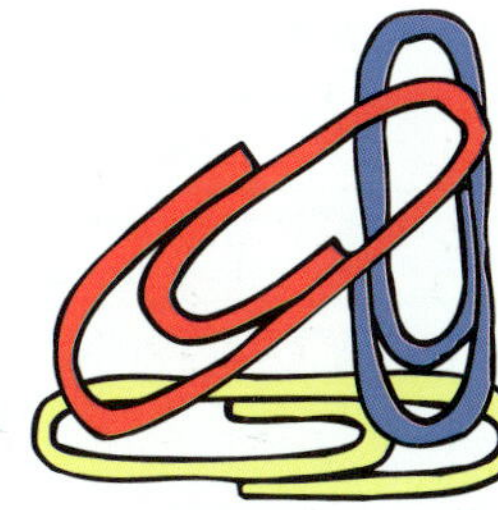

Use a paper clip to measure the length of the green line segment.

The green line segment is ________ paper clips long.

Explain why the green line segment's length is greater in paper clips than it is in toothpicks.

____

Use your shoe to measure the width of the doorway to your bedroom.

The doorway is ________ shoes wide.

Use an adult's shoe to measure the width of the doorway to your bedroom.

The doorway is ________ shoes long.

Explain why the two shoe measurements are not the same.

____

Use the base 10 blocks to find each sum. Write and say each number sentence. Be sure to include an addition sign.

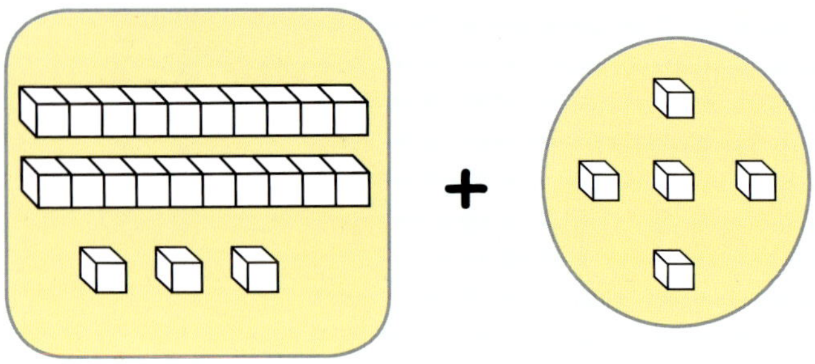

_____ _____ = _____

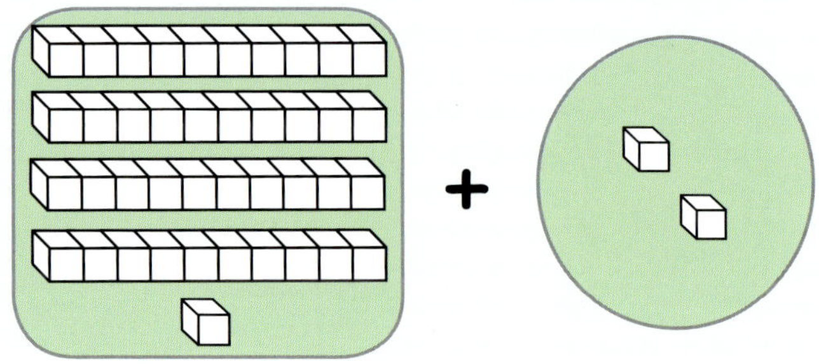

_____ _____ = _____

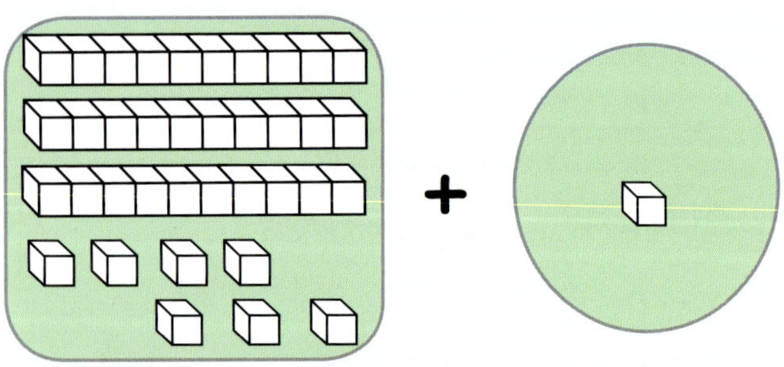

_____ _____ = _____

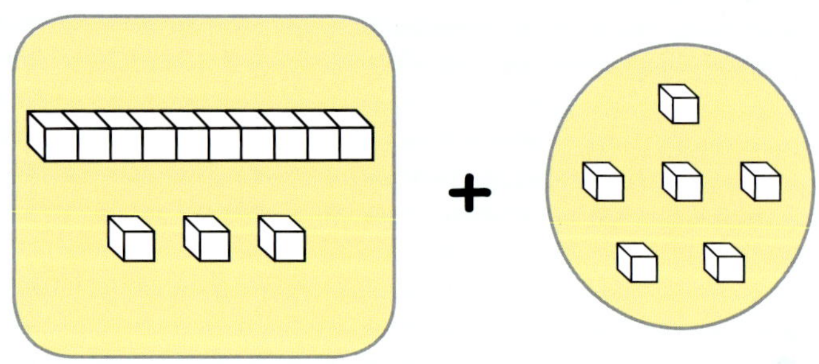

_____ _____ = _____

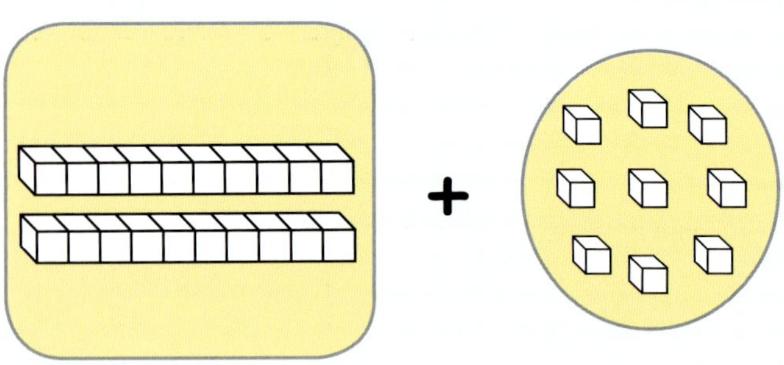

_____ _____ = _____

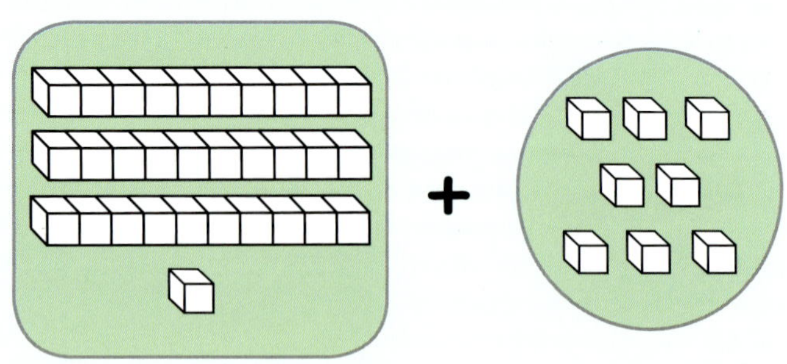

_____ _____ = _____

Touch and say the numerals in the chart, then write the missing numerals.

1	2	3	4	5	6	7	8	9	10
									20
									30
									40
									50
									60
									70
									80
									90
91	92	93	94	95	96	97	98	99	100

Trace the numerals and do the problems. Go in order from left to right. Then, on the next page, connect the answer dots to find the mystery object.

32 + 5 = ______	41 + 2 = ______
53 + 4 = ______	61 + 8 = ______
30 + 9 = ______	71 + 2 = ______
38 + 1 = ______	92 + 7 = ______
60 + 5 = ______	40 + 2 = ______
30 + 7 = ______	50 + 9 = ______
72 + 2 = ______	11 + 2 = ______
22 + 2 = ______	21 + 1 = ______
36 + 3 = ______	53 + 2 = ______
82 + 5 = ______	42 + 3 = ______

I love the wind and it makes me go fast. I do not have any wheels, but I do have a steering wheel.

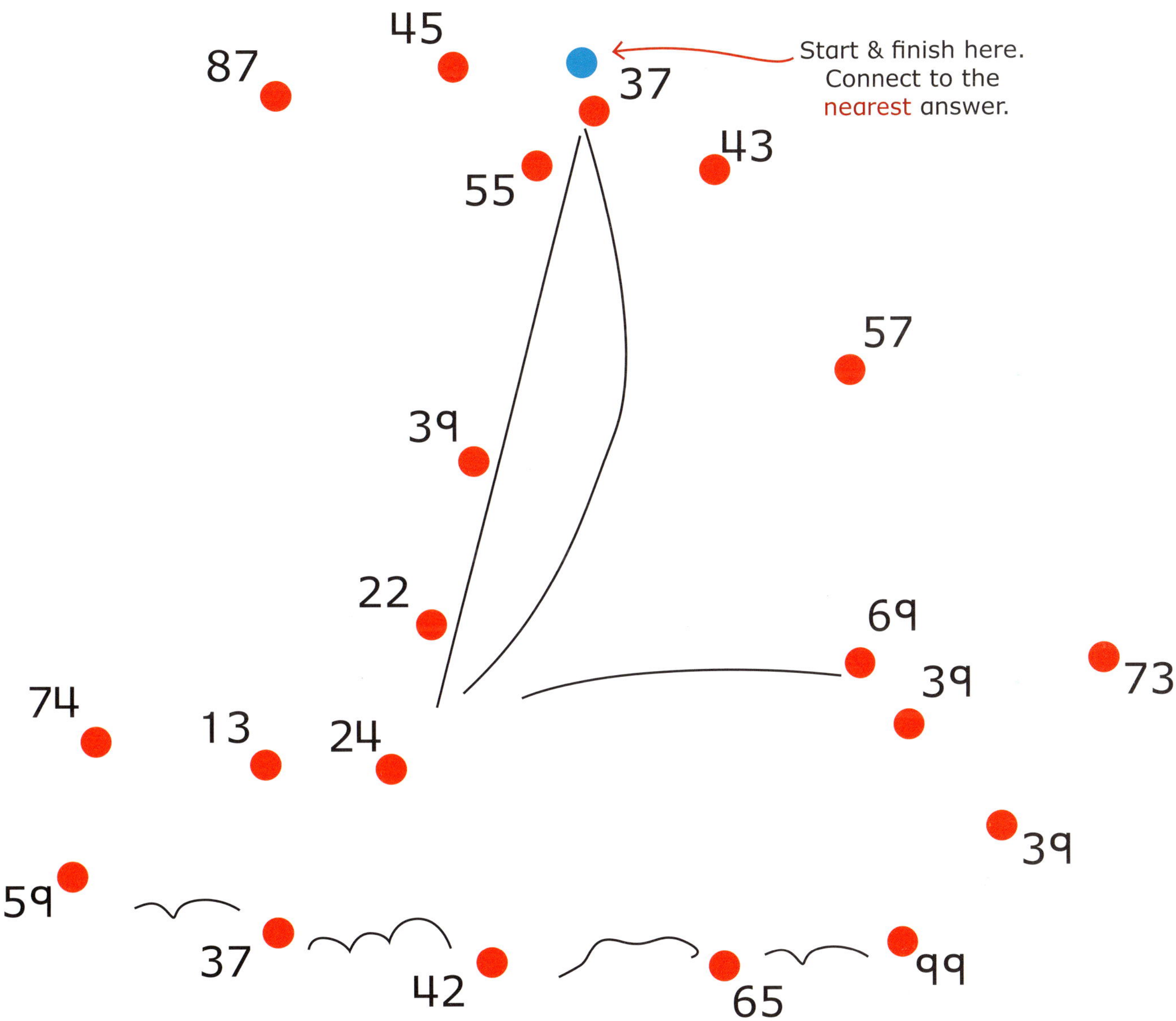

Color the mystery object and add something new to the picture.

Draw and color the next three shapes that would continue the pattern in each row.

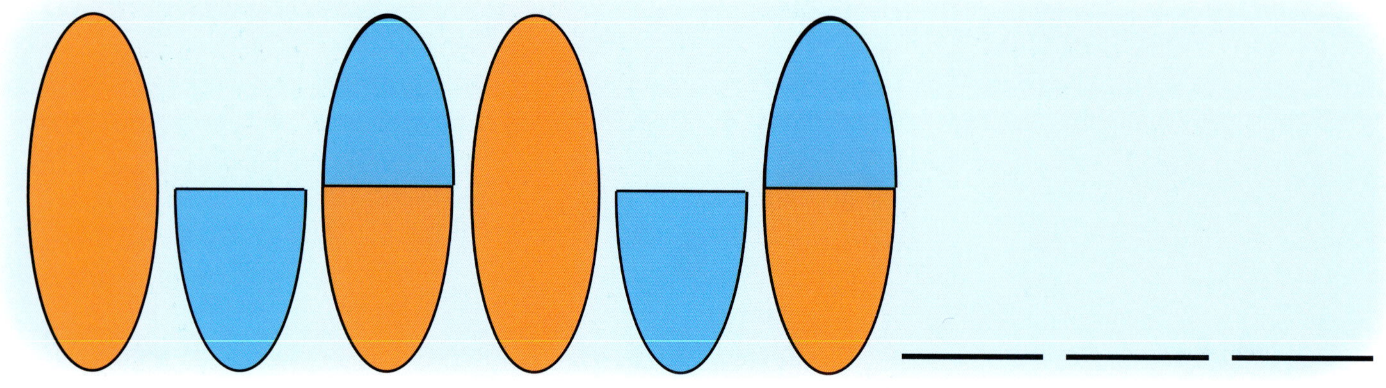

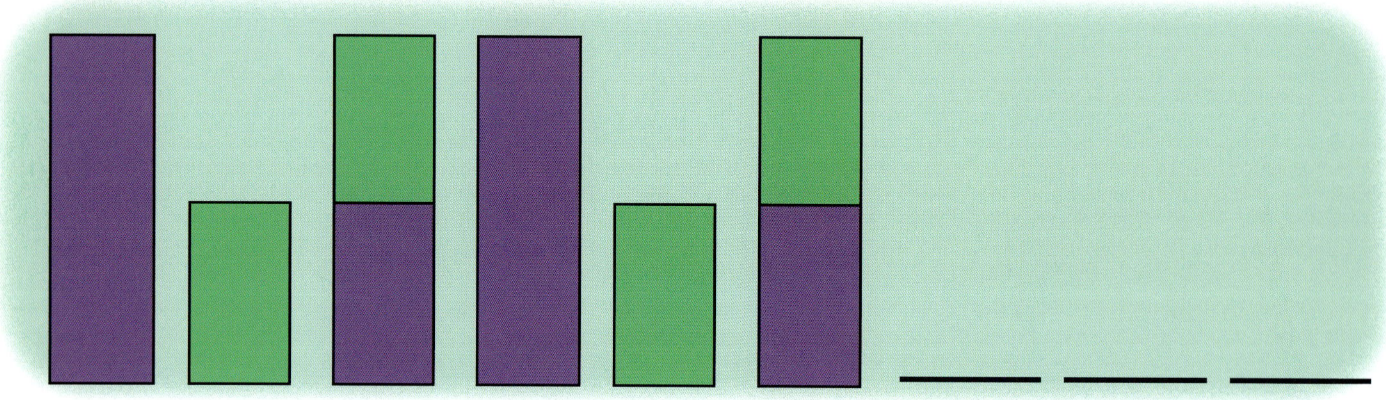

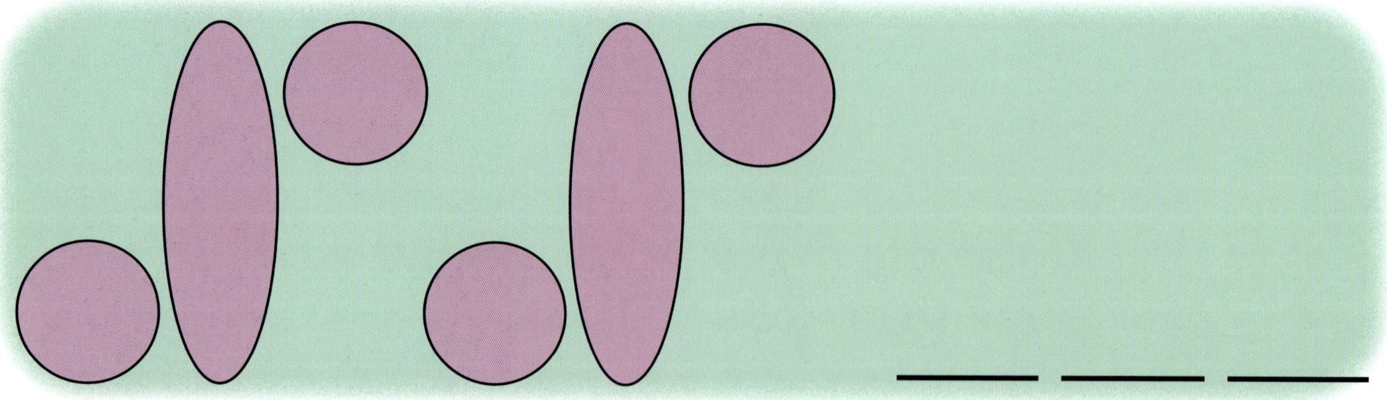

Show the Times

Show the time you eat breakfast.

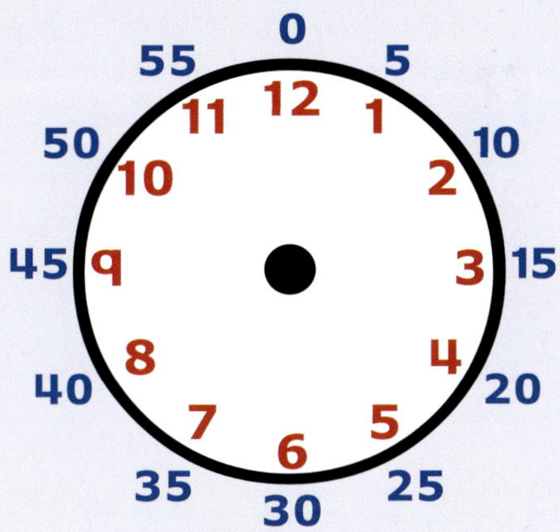

Show the time you go to see a movie.

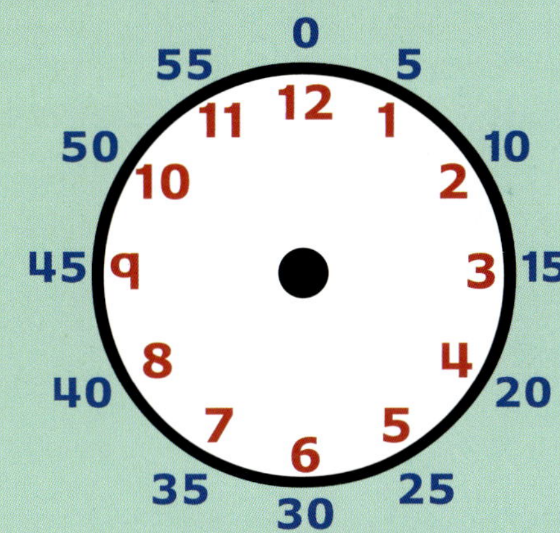

Show the time you take a bath.

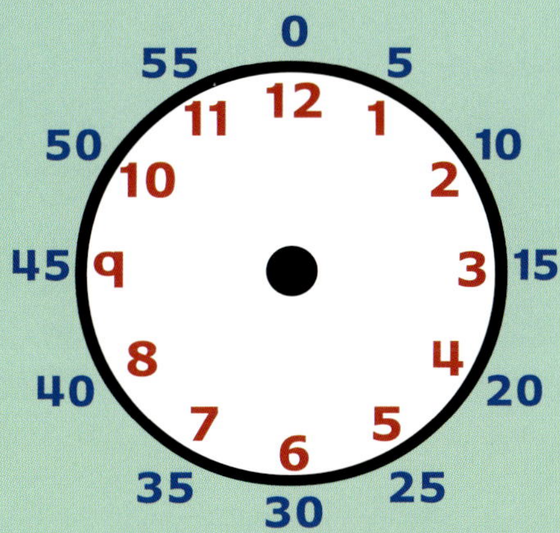

Show the time you get out of school.

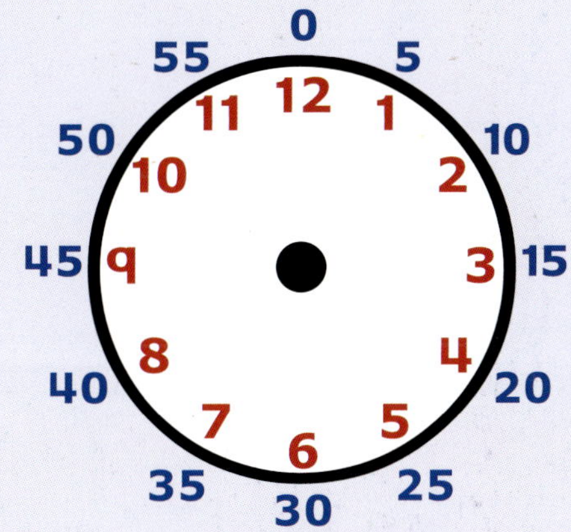

Trace the numerals and do the problems. Go in order from left to right. Then, on the next page, connect the answer dots to find the mystery object.

70 - 20 = ____	90 - 80 = ____
80 - 70 = ____	50 - 20 = ____
80 - 40 = ____	90 - 30 = ____
60 - 20 = ____	70 - 60 = ____
30 - 20 = ____	50 - 10 = ____
40 - 20 = ____	90 - 80 = ____
60 - 50 = ____	50 - 30 = ____
80 - 50 = ____	70 - 30 = ____
80 - 70 = ____	90 - 50 = ____
90 - 40 = ____	40 - 10 = ____

I sit high so you can see me in a crowd. I am good for protection from the rain and the sunshine.

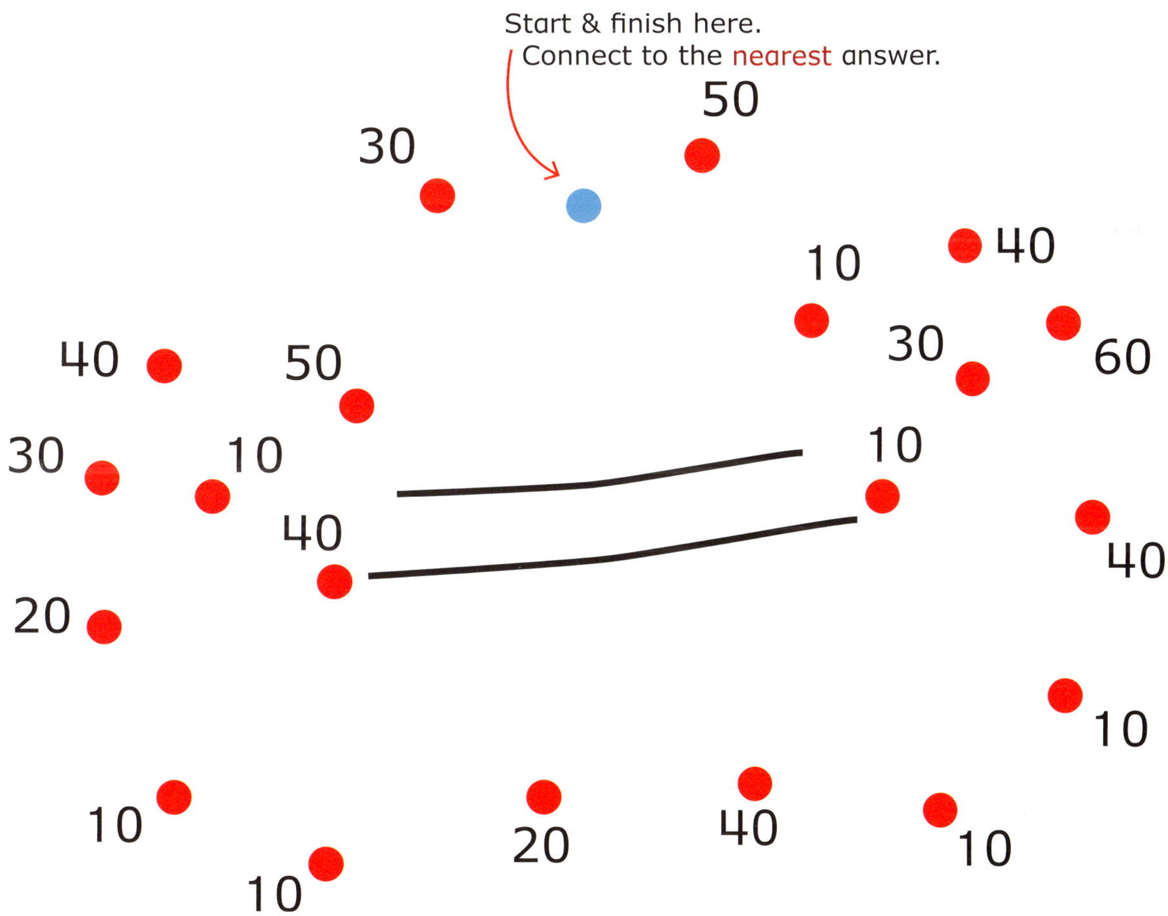

Color the mystery object and add something new to the picture.

Draw a line of symmetry on each figure.

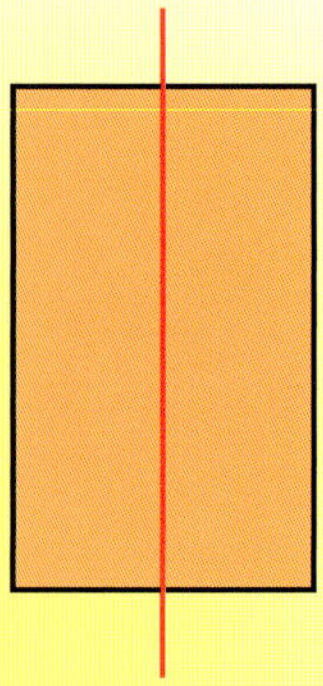

EXAMPLE

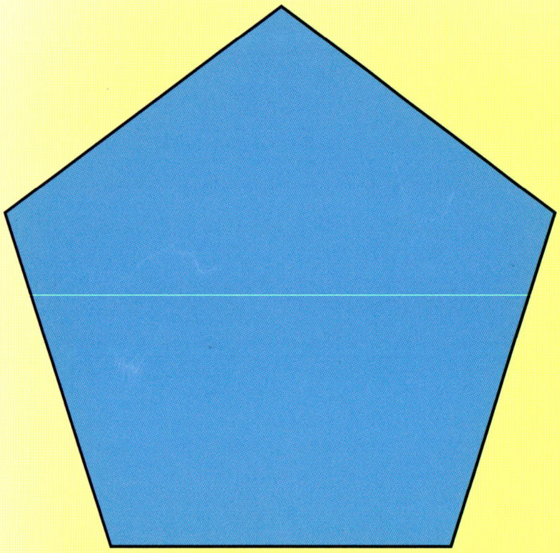

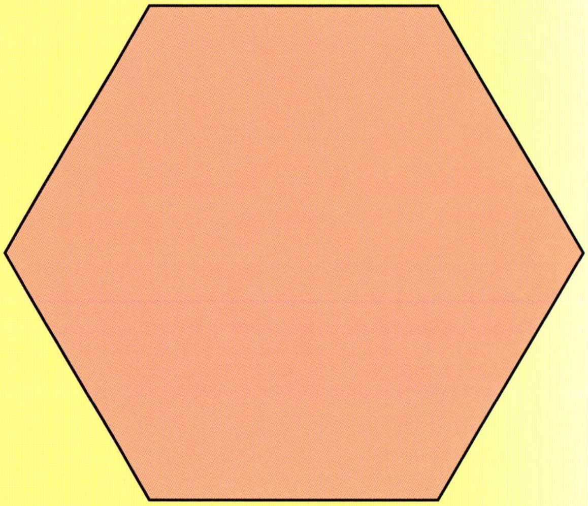

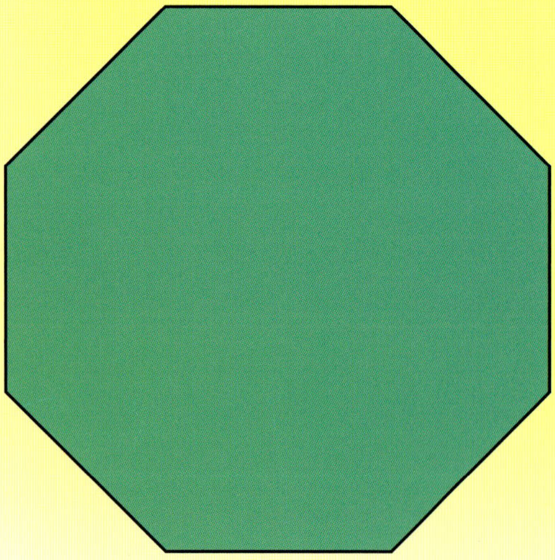

You have a dime, 2 nickels, and 3 pennies. What is the total amount of money you have?

Two is greater than one. Would you trade 2 quarters for 1 dollar?

Two is greater than one. Would you trade 2 nickels for 1 quarter?

You start with a nickel. Each of the next 8 people you meet gives you 1¢. Then you find another nickel. What is the total amount of money you have then?

You get 5¢ for each book you read. If you read 4 books, what is the fewest number of coins you can use to show your total amount of money? What are those coins?

Three is more than 2, so would you trade 3 nickels for 2 dimes? Explain your answer.

Two is more than 1, so would you trade 2 dimes for 1 quarter? Explain your answer.

Two is more than 1, so would you trade 2 dimes for 1 nickel? Explain your answer.

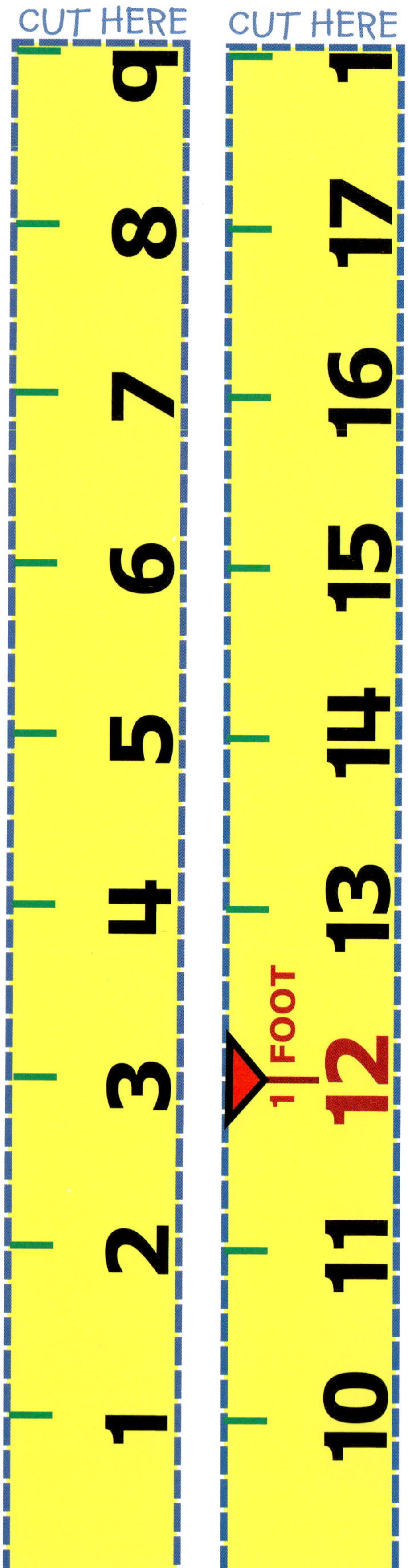

Use a measuring tape or cut out and tape this ruler together to measure the objects on the next page. Do not overlap the pieces of the ruler. Lay them end to end.

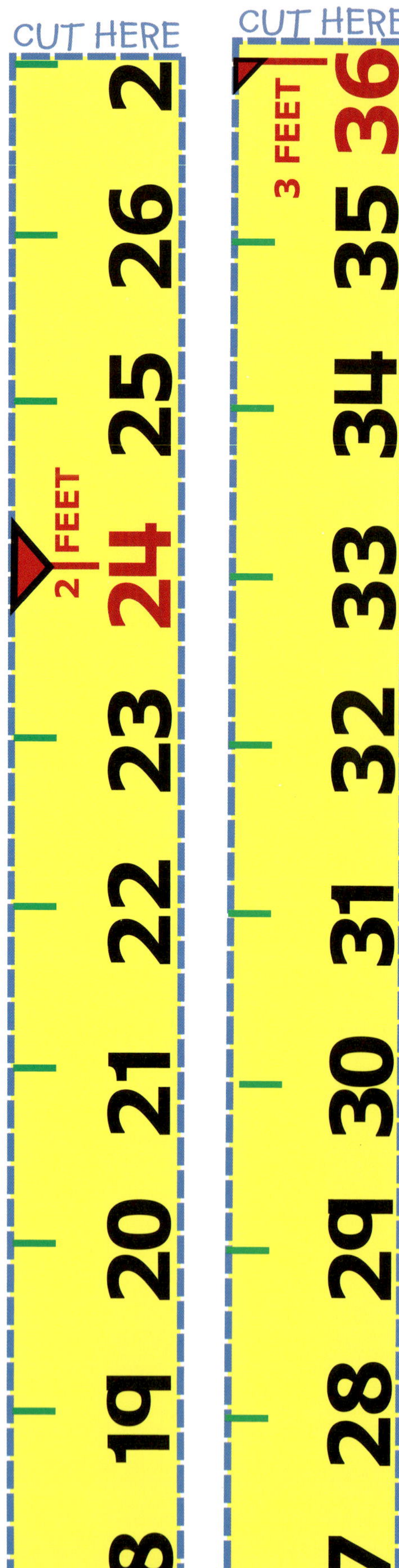

What is the width of the doorway to your bedroom?

_______FEET_______INCHES

What is the height of the top of the doorknob of your bedroom door from the floor?

_______FEET_______INCHES

How long is the shoe you are wearing?

_______FEET_______INCHES

How long is your arm?

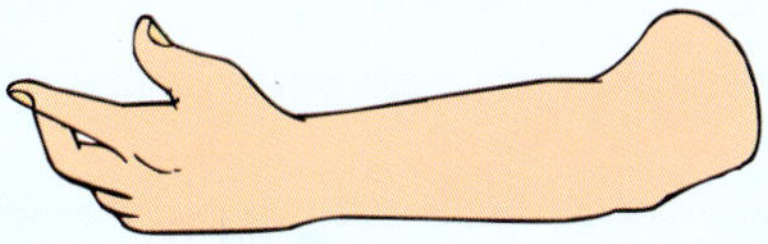

_______FEET_______INCHES

How long is your mom's arm?

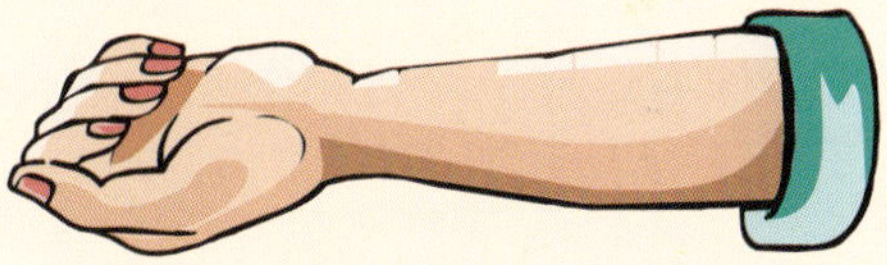

_______FEET_______INCHES

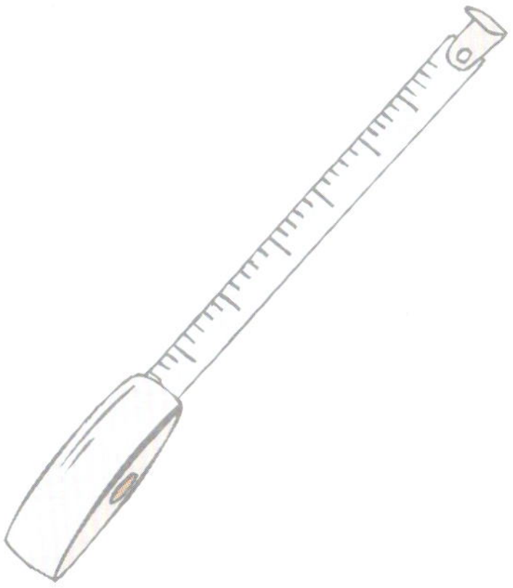

How tall is your cereal box?

_______FEET_______INCHES

Trace the numerals and do the problems. Go in order from left to right. Then, on the next page, connect the answer dots to find the mystery object.

30 + 50 = ______ 40 + 20 = ______

50 + 10 = ______ 60 + 10 = ______

30 + 40 = ______ 70 + 20 = ______

30 + 10 = ______ 90 + 0 = ______

60 + 10 = ______ 40 + 20 = ______

30 + 60 = ______ 50 + 40 = ______

70 + 20 = ______ 10 + 20 = ______

20 + 20 = ______ 20 + 10 = ______

30 + 30 = ______ 50 + 20 = ______

20 + 30 = ______ 40 + 30 = ______

I love flowers. I like to sit on their petals and sip the nectar.

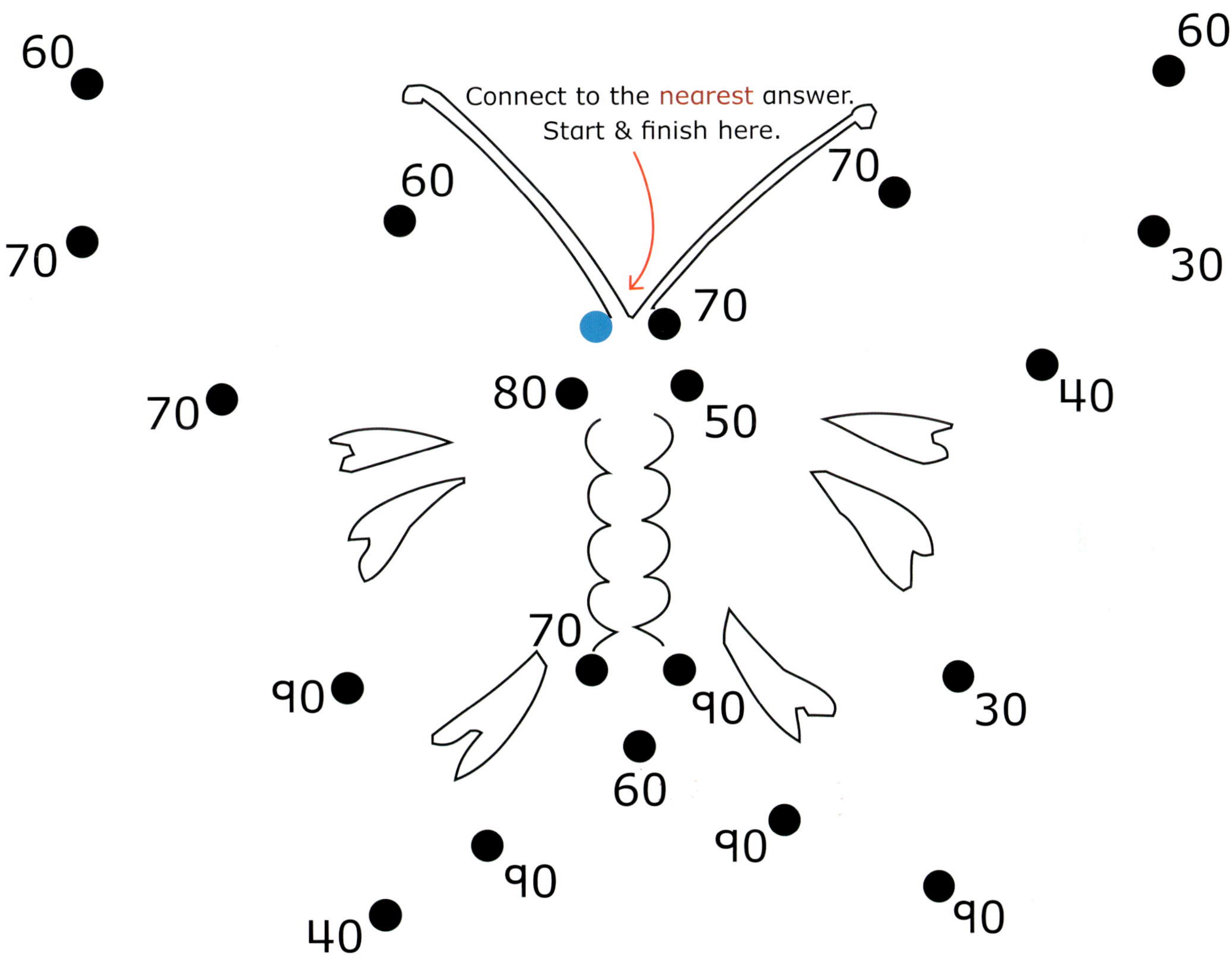

Color the mystery animal and add something new to the picture.

Use the base 10 blocks to solve each problem. Then write and say each number sentence. Be sure to include an addition sign.

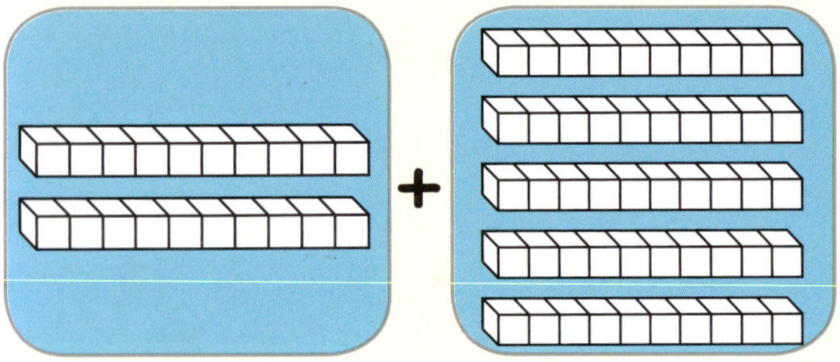

_____ _____ = _____

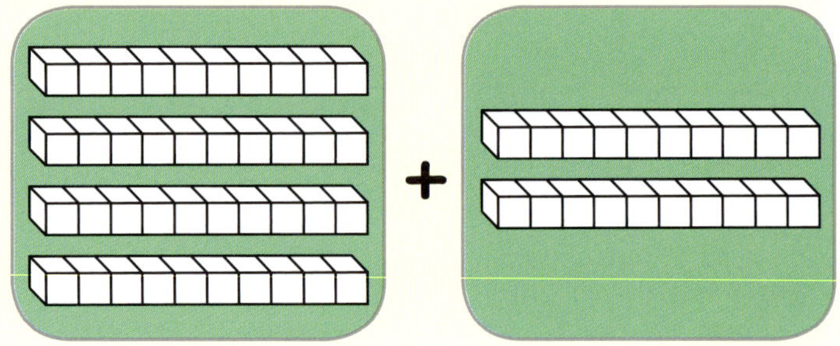

_____ _____ = _____

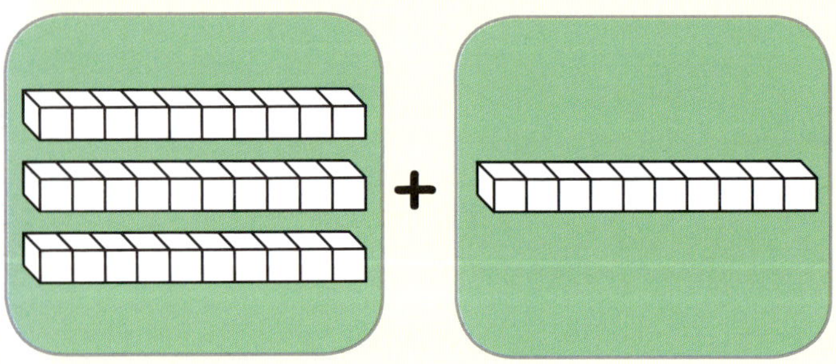

_____ _____ = _____

_____ _____ = _____

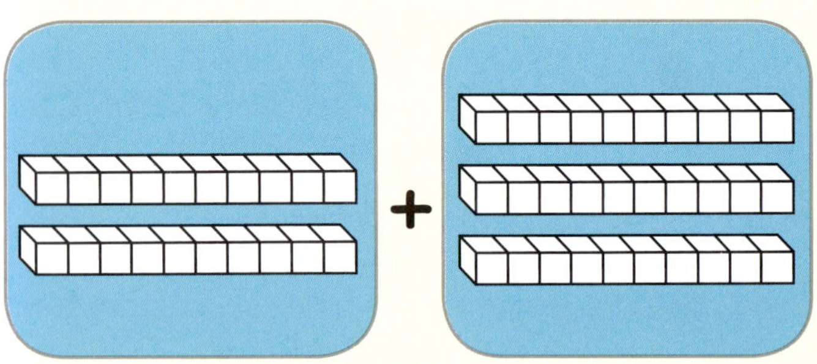

_____ _____ = _____

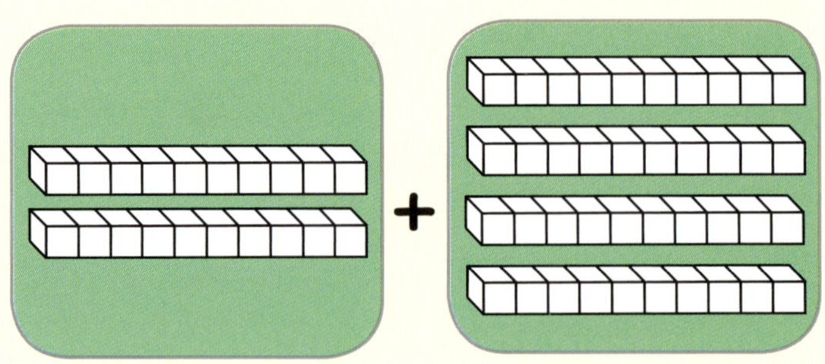

_____ _____ = _____

Use the pictures of the yardstick and the 12 inch ruler to answer the questions.

Picture of a yardstick (3 feet = 1 yard)

picture of a 12 inch (1 foot) ruler

Which is shorter, 11 inches or 9 inches?

Which is shorter, 2 feet or 1 yard?

Which is longer, 27 inches or 1 yard?

Which is longer, 27 inches or 2 feet?

Which is longer, 36 inches or 1 yard?

Which is shorter, 19 inches or 3 feet?

Circle the shape each object in a blue circle will make when it is folded.

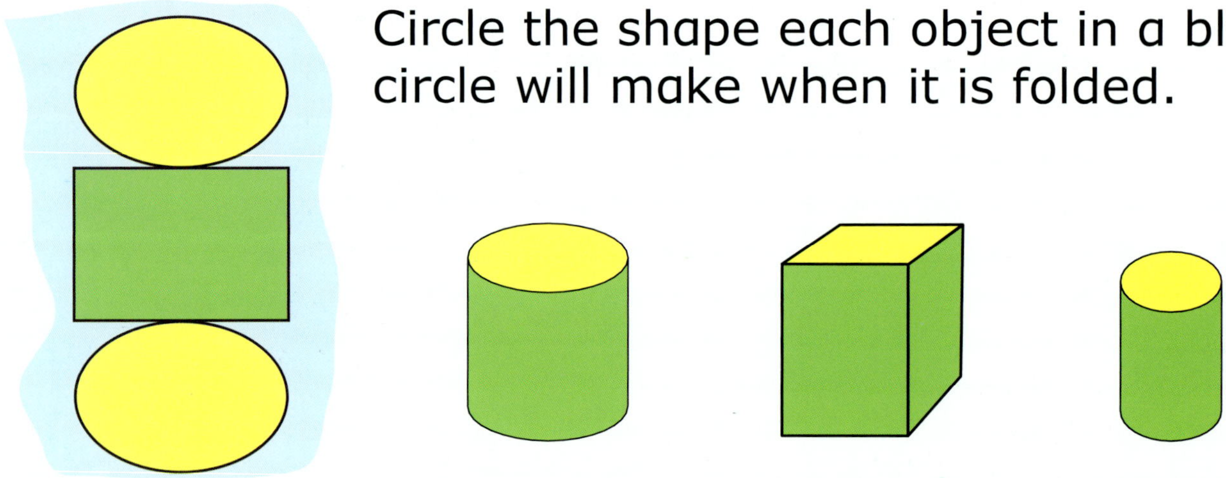

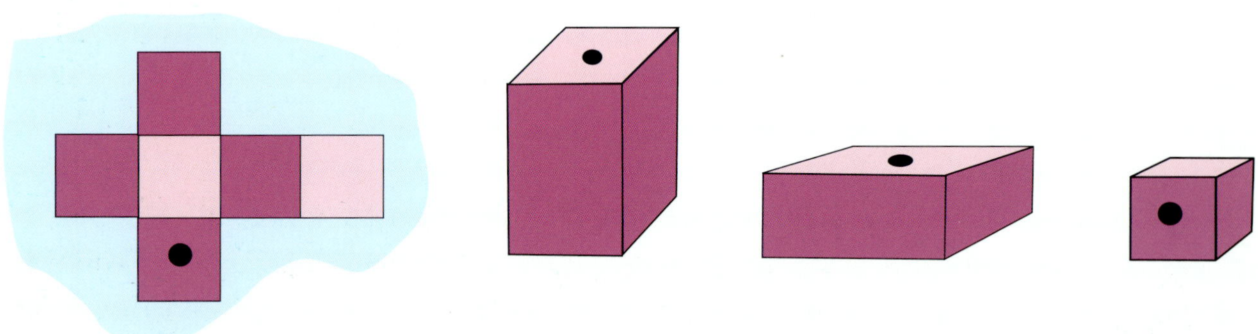

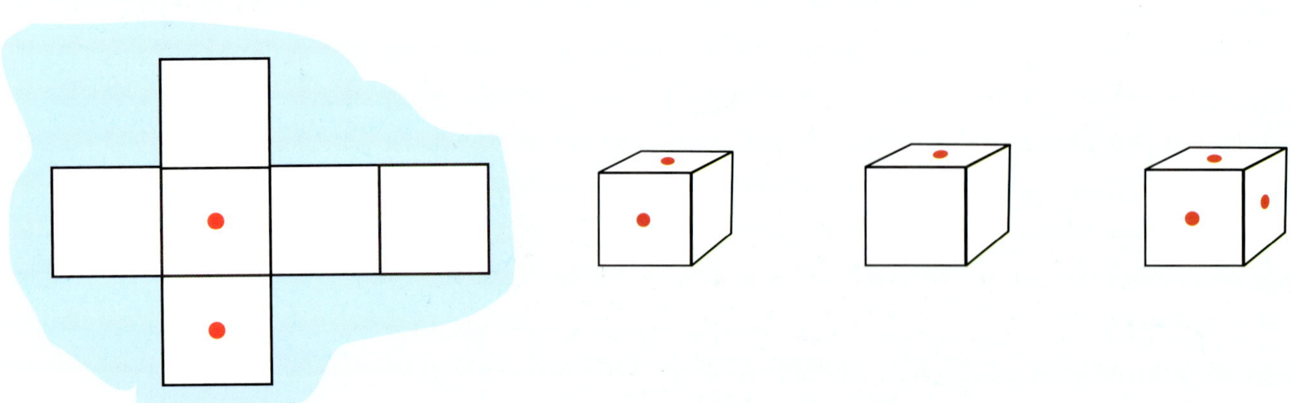

Draw a line segment to connect the objects of the same length.

Circle the coins you would use to make the amount in the blue circle.

22¢ =

25¢ =

19¢ =

21¢ =

Individually add the ones and tens columns.

Add the following numbers by adding the **green ones** column and then the **blue tens** column.

Tens Column
Ones Column

12 +11 23	14 +31	52 +11
81 +18	36 +32	17 +21

Add the following numbers by adding the ones column and then the tens column.

34 +22	61 +25	58 +11
18 +11	66 +33	87 +12

Do the problems. Go in order from left to right. Then, on the next page, connect the answer dots to find the mystery animal.

13 +14	23 +41	36 +22	17 +21
11 +52	42 +24	33 +24	21 +11
14 +30	33 +22	41 +36	33 +46
23 +25	26 +31	34 +31	35 +12
31 +34	42 +10	22 +22	14 +44

I don't run, but I do hop. Eating vegetables is one of my favorite hobbies.

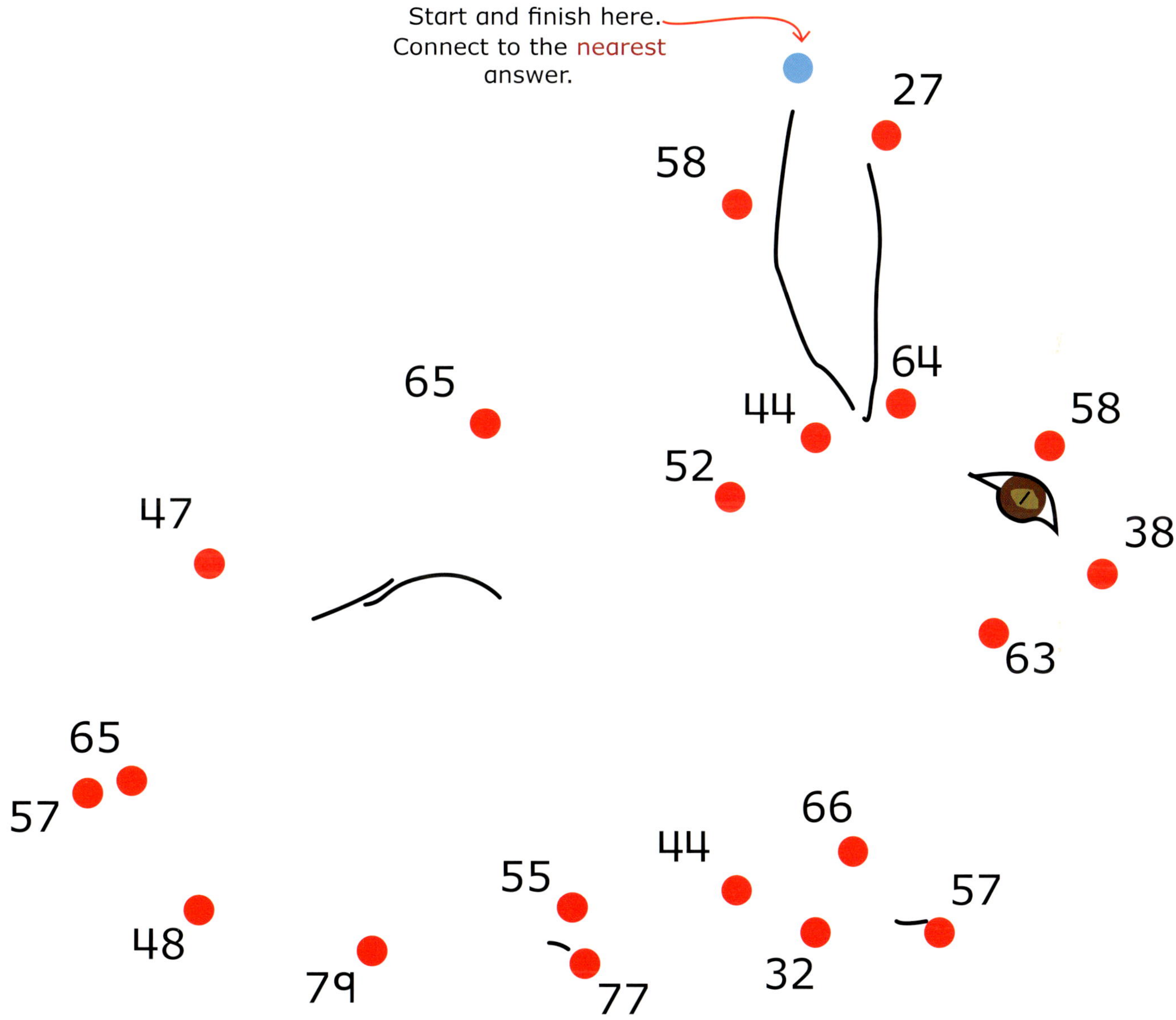

Color the mystery animal and add something new to the picture.

Draw a fish that is more than 5 inches long. Then draw another fish that is 3 inches longer than the first fish you drew.

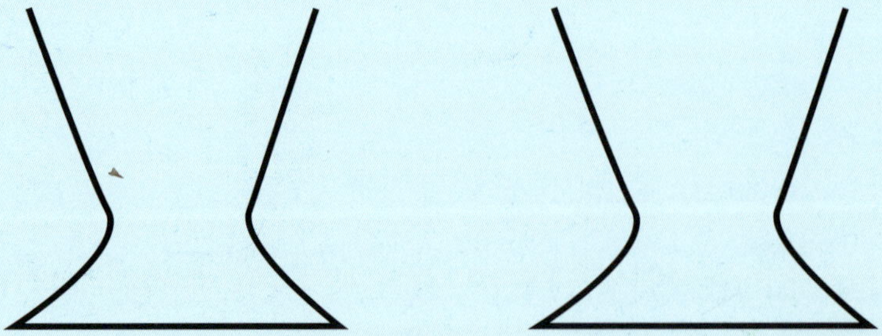

Each of these objects has more than one line of symmetry. Draw each line of symmetry on each figure, then write the number of lines of symmetry each figure has.

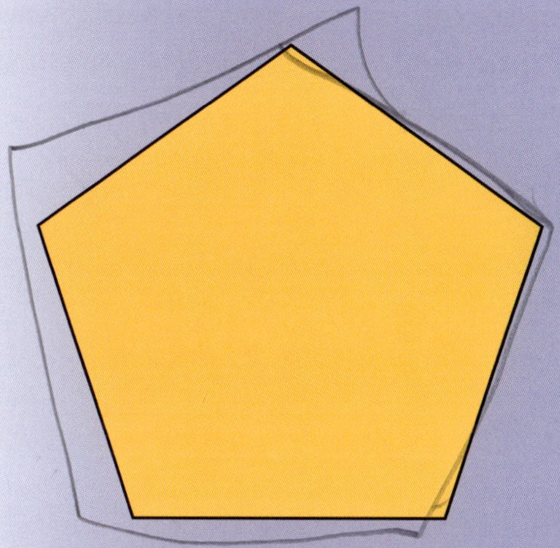

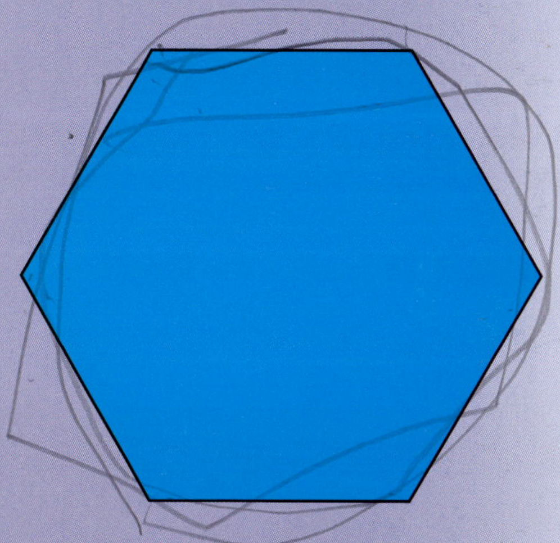

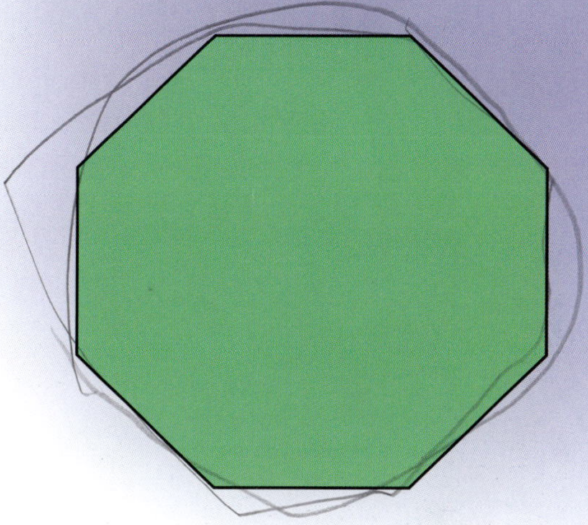

Circle the amount of cents the boy would rather have, 3 pennies or a nickel.

Circle whether the girl is probably going to be eating lunch or going to bed.

lunch   bed

Circle whether the children will probably be warm or cold.

warm   cold

The number 21 is between 20 and 22.

20 21 22

Trace each pair of numerals, then write the numeral that comes between them.

22	___	24	42	___	44
31	___	33	45	___	47
19	___	21	17	___	19
51	___	53	65	___	67
61	___	63	15	___	17
68	___	70	71	___	73
74	___	76	79	___	81
80	___	82	90	___	92
95	___	97	98	___	100

Match each set of base ten blocks with the correct quantity in pink.

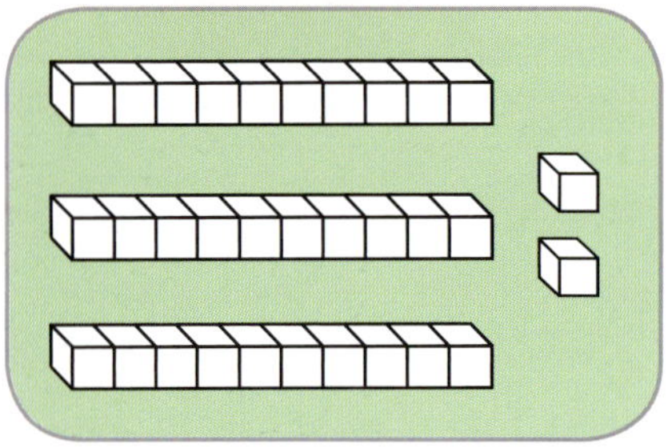

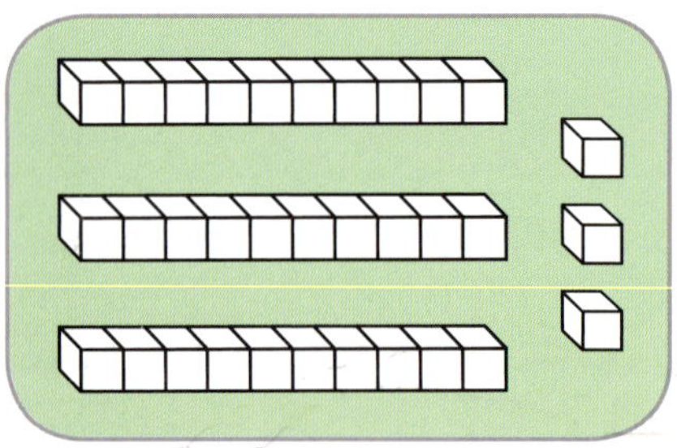

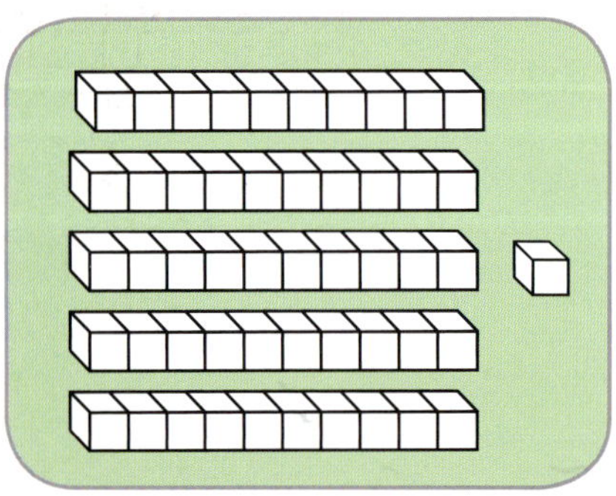

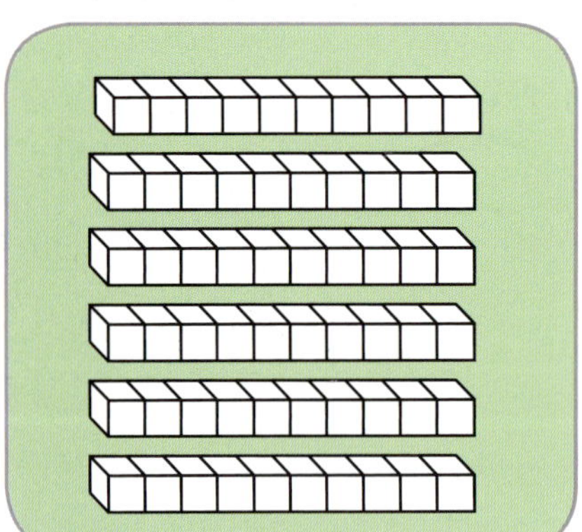

33

44

32

28

51

60

85

43

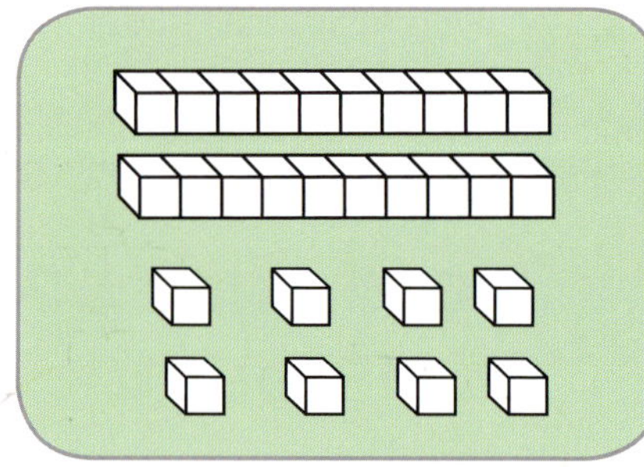

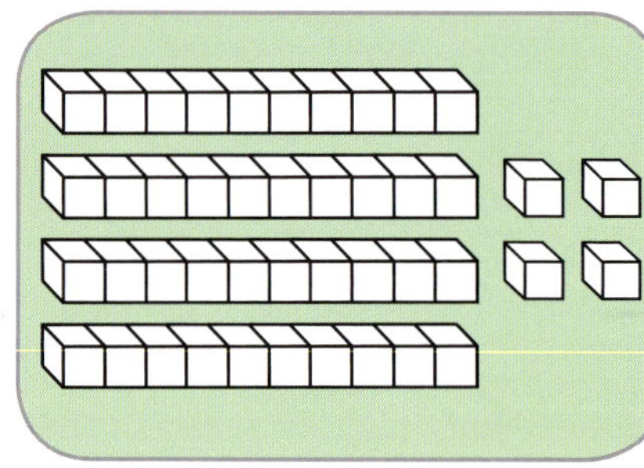

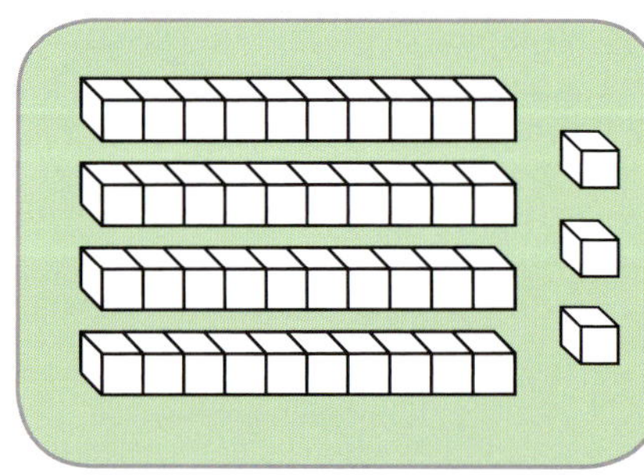

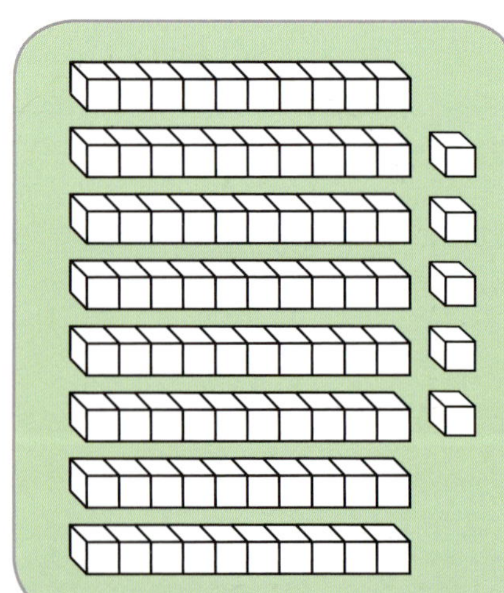

Use the base 10 blocks to solve each problem.
Then write and say each number sentence.
Be sure to include an addition sign.

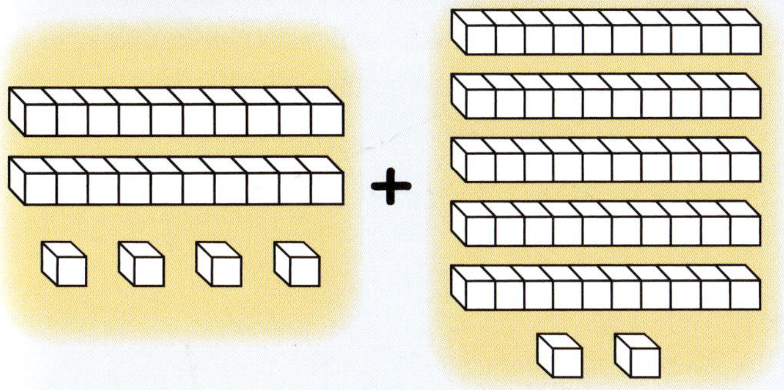

______ ______ = ______

______ ______ = ______

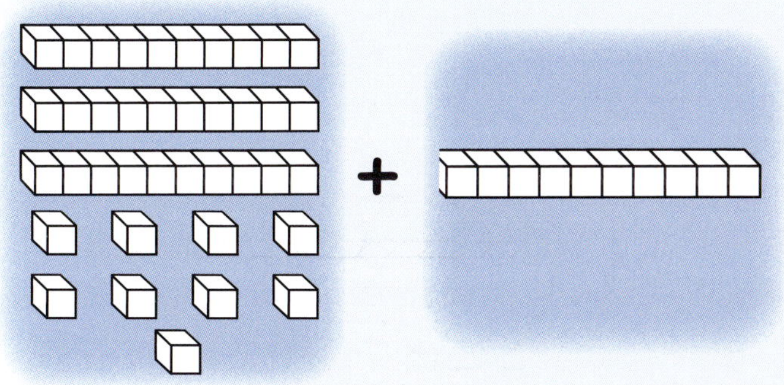

______ ______ = ______

______ ______ = ______

______ ______ = ______

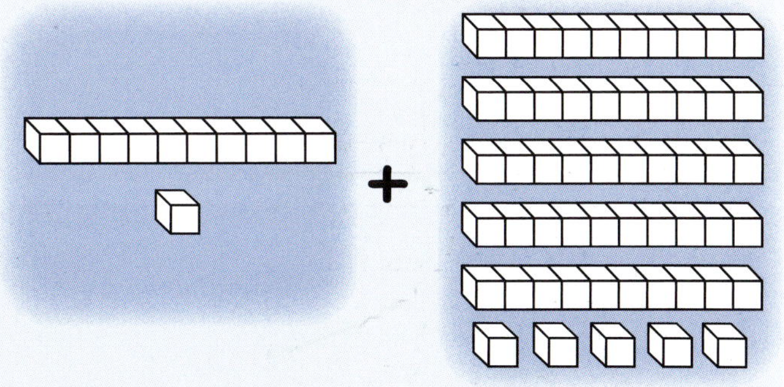

______ ______ = ______

Each of these objects has more than one line of symmetry. Draw each line of symmetry on each figure, then write the number of lines of symmetry each figure has.

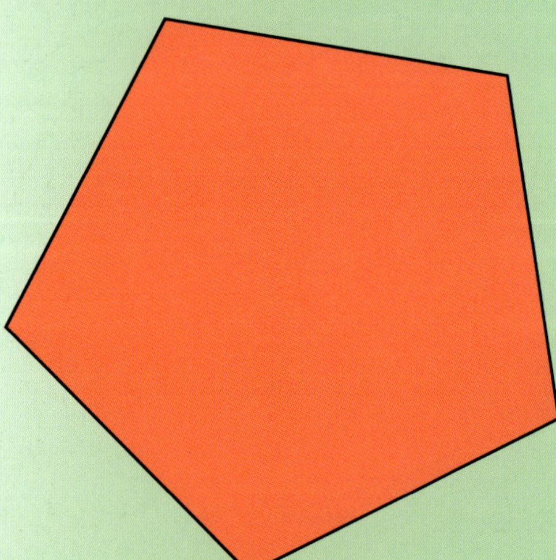

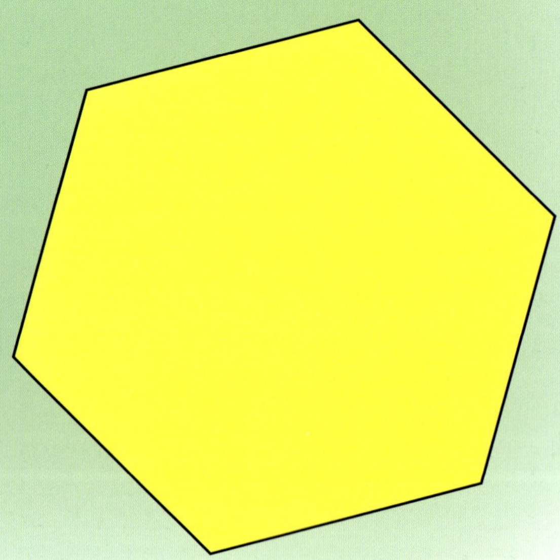

Color or write the missing fraction.

Example

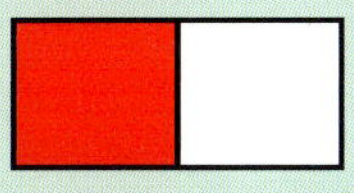

$\frac{1}{2}$

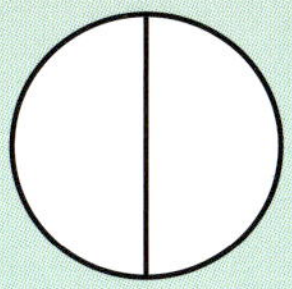

$\frac{1}{2}$

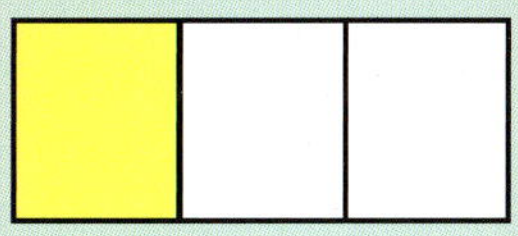

$\frac{1}{}$

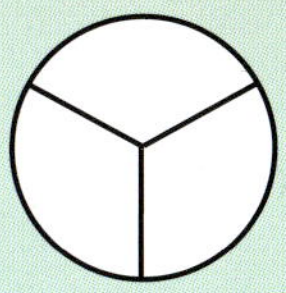

$\frac{1}{3}$

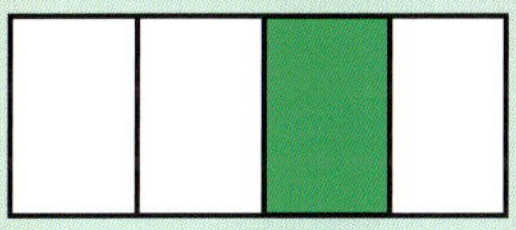

$\frac{1}{}$

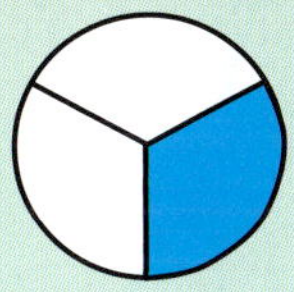

$\frac{1}{}$

$\frac{1}{4}$

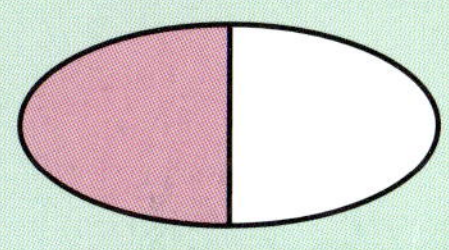

$\frac{1}{}$

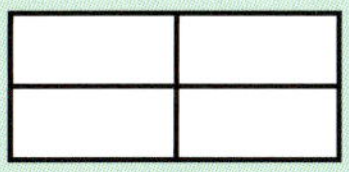

$\frac{1}{4}$

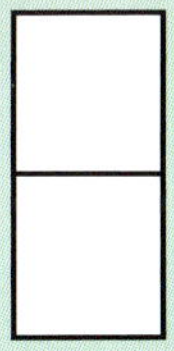

$\frac{1}{2}$

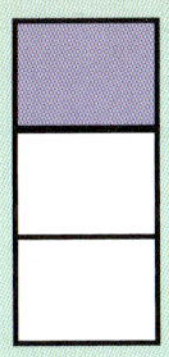

$\frac{1}{}$

$\frac{1}{}$

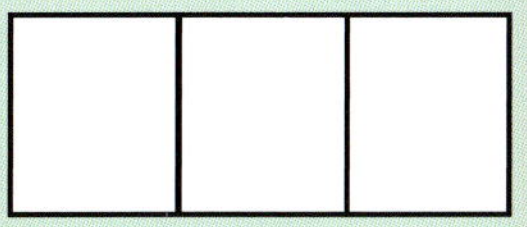

$\frac{1}{3}$

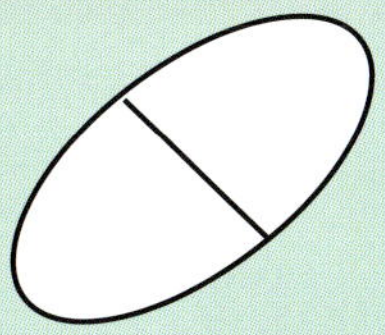

$\frac{1}{2}$

$\frac{1}{}$

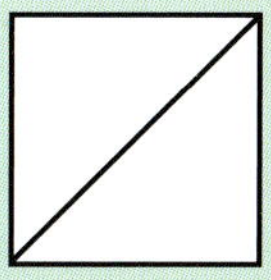

$\frac{1}{2}$

Addition is +

Do the problems. Go in order from left to right. Then, on the next page, connect the answer dots to find the mystery object.

23 +24	43 +41	36 +52	77 +21
41 +52	42 +54	73 +24	41 +41
54 +32	83 +12	61 +36	51 +46
63 +25	66 +31	44 +33	45 +54
61 +34	92 + 6	42 +56	34 +35

I have keys but no lock. I can't walk, but I travel many places.

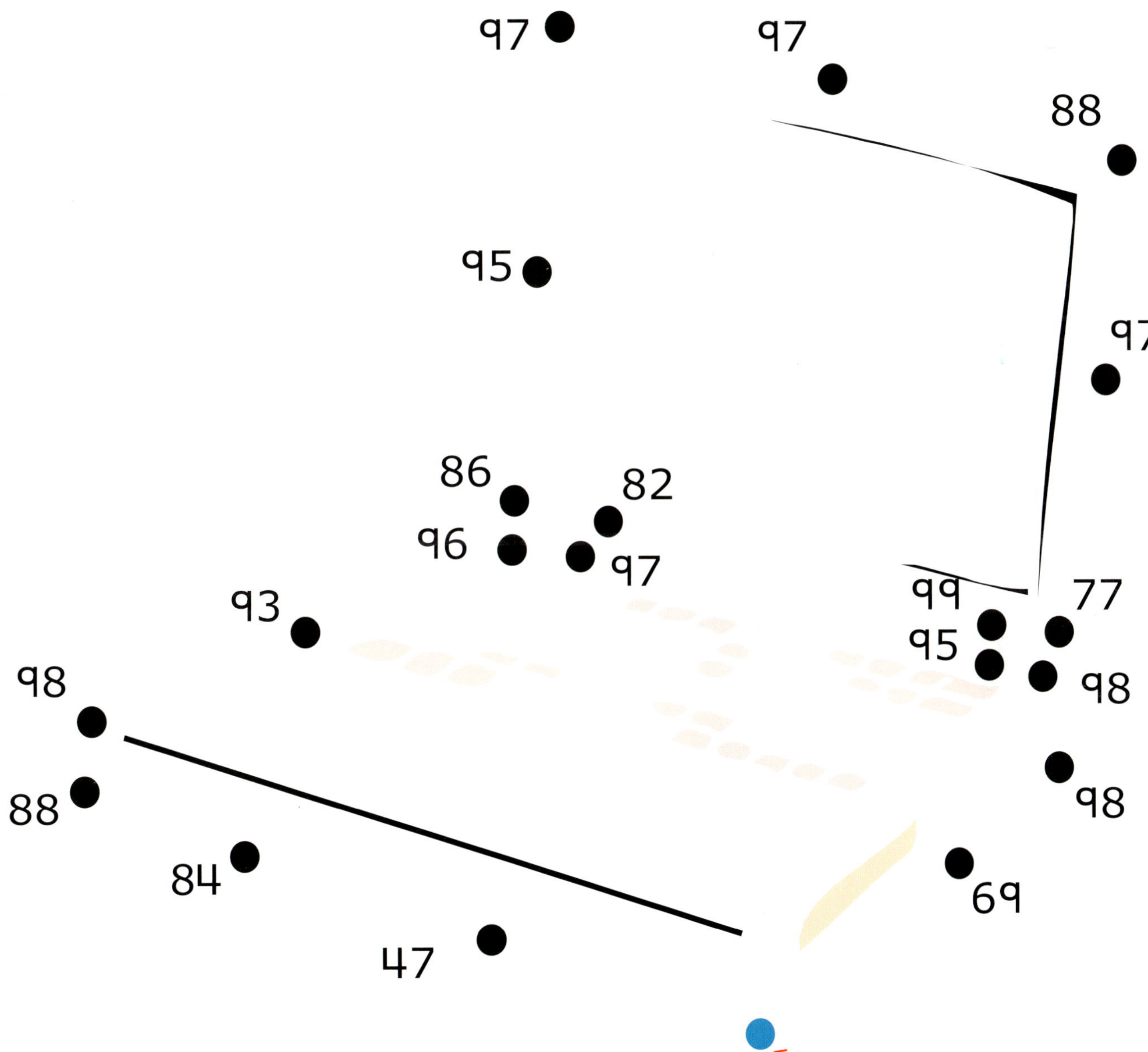

Start & finish here.
Connect to the nearest answer.

Color the mystery object and add something new to the picture.

Use the base 10 blocks to solve each problem. Then write and say each number sentence.

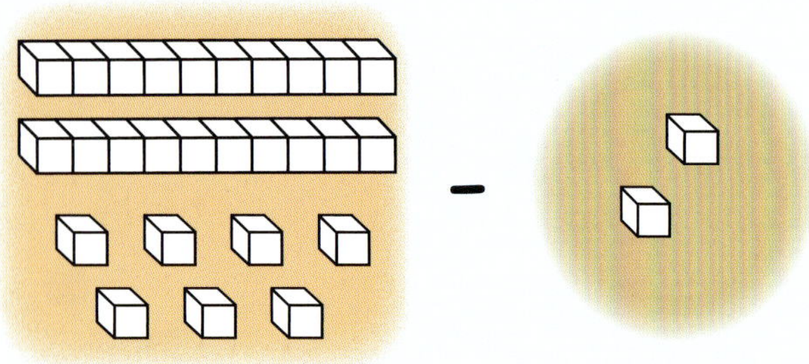

______ ______ = ______

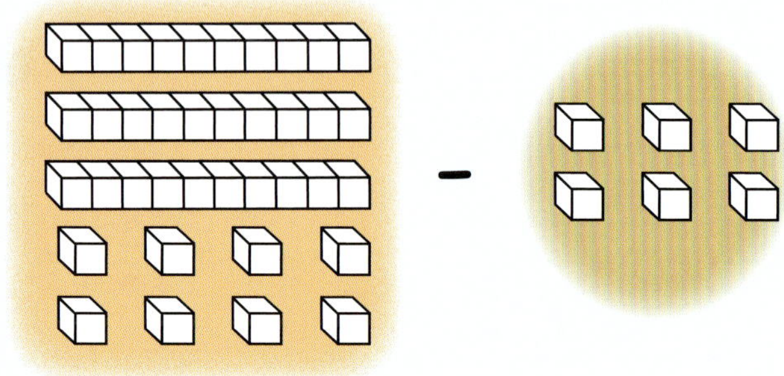

______ ______ = ______

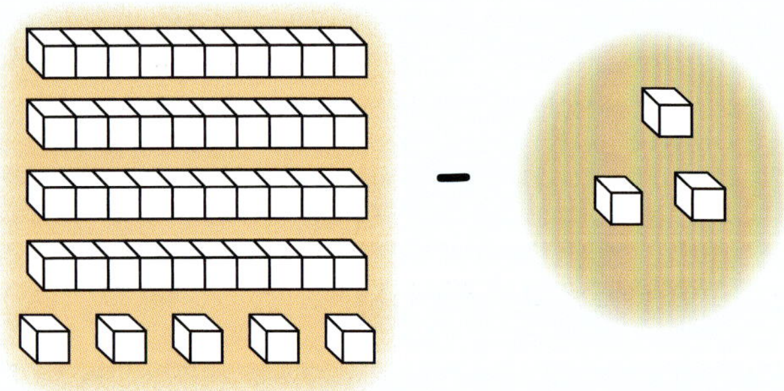

______ ______ = ______

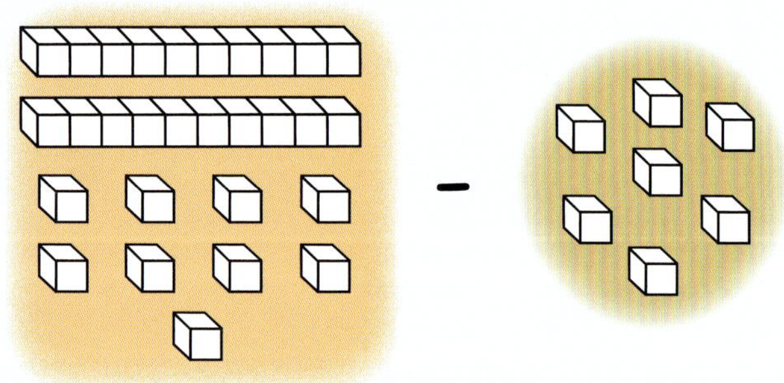

______ ______ = ______

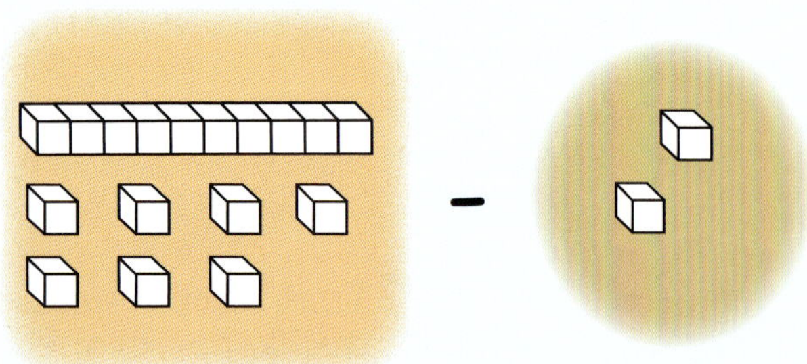

______ ______ = ______

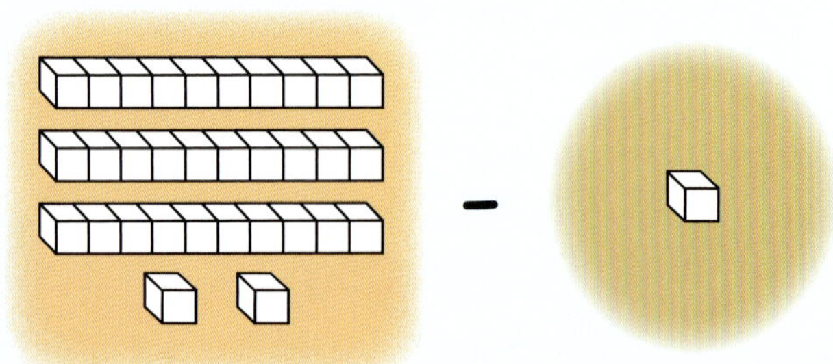

______ ______ = ______

Individually subtract the ones and tens columns.

Subtract the following numbers by subtracting the **green ones** column and then the **blue tens** column.

Tens Column — Ones Column

```
  14      23      34
-  1    -  2    -  3
----    ----    ----
  13

  76      47      95
-  2    -  6    -  2
----    ----    ----
```

Subtract the following numbers by subtracting the ones column and then the tens column.

```
  73      23      33
-  1    -  2    -  3
----    ----    ----

  68      84      44
-  1    -  3    -  4
----    ----    ----
```

Do the problems. Go in order from left to right. Then, on the next page, connect the answer dots to find the mystery object.

28 - 5	45 - 3	23 - 1	25 - 4
43 - 2	34 - 3	86 - 1	77 - 4
35 - 2	58 - 4	46 - 3	93 - 2
27 - 6	39 - 9	56 - 5	49 - 8
68 - 7	99 - 0	89 - 7	46 - 5

I have buttons but no button holes. You can hear me before you can see me.

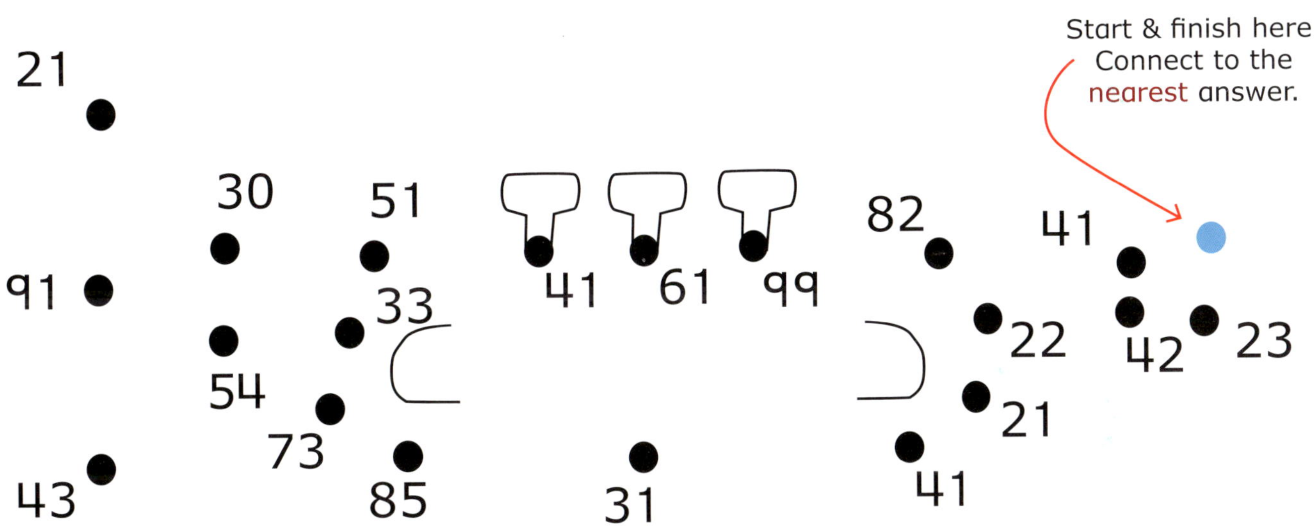

Color the mystery object and add something new to the picture.

You have 10 green leaves. If you put them in an empty bag, closed your eyes, and picked only one leaf, what color would it be?

You have 10 red and 2 yellow leaves. If you put them in an empty bag, closed your eyes, and picked only one leaf, what color would it probably be?

You have 8 brown and 2 orange leaves. If you put them in an empty bag, closed your eyes, and picked only one leaf, what color would it probably be?

You have 6 green, 3 purple, and 3 red leaves. If you put them in an empty bag, closed your eyes, and picked only one leaf, what color would it probably be?

You have 7 orange and 5 yellow leaves. If you put them in an empty bag, closed your eyes, and picked only one leaf, what color would it probably be?

Each of these objects has more than one line of symmetry. Draw each line of symmetry on each figure, then write the number of lines of symmetry each figure has.

____

____

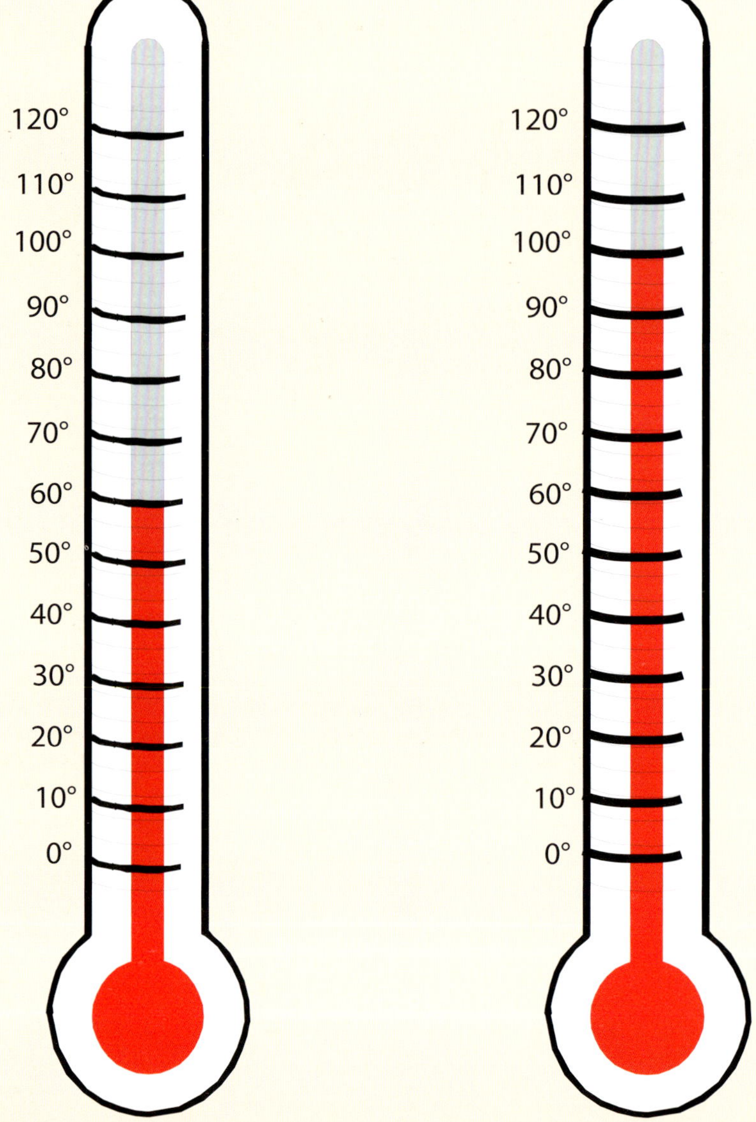

Circle the thermometer that shows the hotter day.

What is the temperature of the hotter day? __________

Explain your answer.

What is the temperature of the cooler day? __________

Explain your answer.

Draw a line of symmetry on each letter. Some letters have more than one line of symmetry.

D H O

Which letter has the most lines of symmetry?___________

Which letter has the fewest lines of symmetry?___________

Use the lines of symmetry to draw two symmetric squares.

EXAMPLE

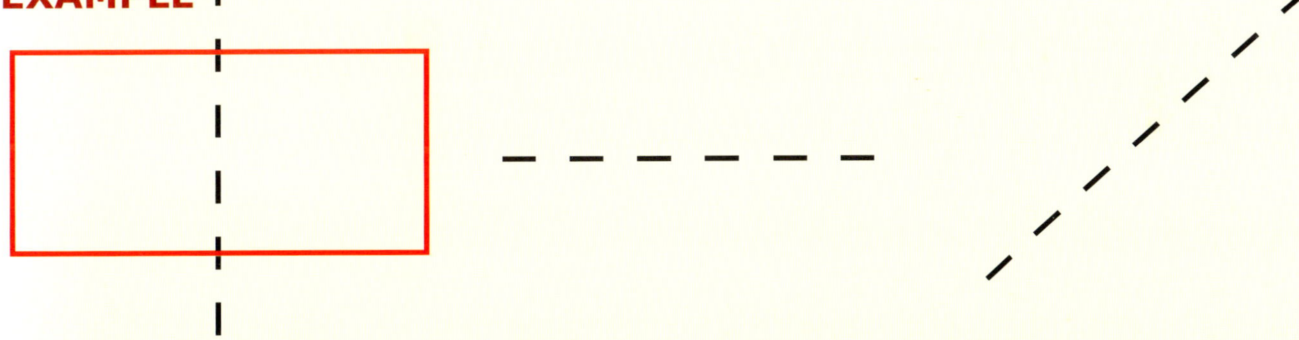

Use the lines of symmetry to draw two symmetric triangles.

The number 98 comes before 99.

98 **99**

Trace each numeral, then write the numeral that comes before it.

_____ 76	_____ 24
_____ 32	_____ 95
_____ 100	_____ 47
_____ 49	_____ 71
_____ 86	_____ 40
_____ 59	_____ 54
_____ 99	_____ 82

Tim, Jan, Blackie, and Boots are standing between two 10 foot poles. Jan is exactly 5 feet tall. Circle your answers and explain your thinking.

Estimate Tim's height.	9 feet	6 feet	4 feet
Estimate Blackie's height.	2 feet	5 feet	4 feet
Estimate Boots's height.	3 feet	2 feet	1 foot
Estimate the height of the fishing pole.	6 feet	7 feet	9 feet

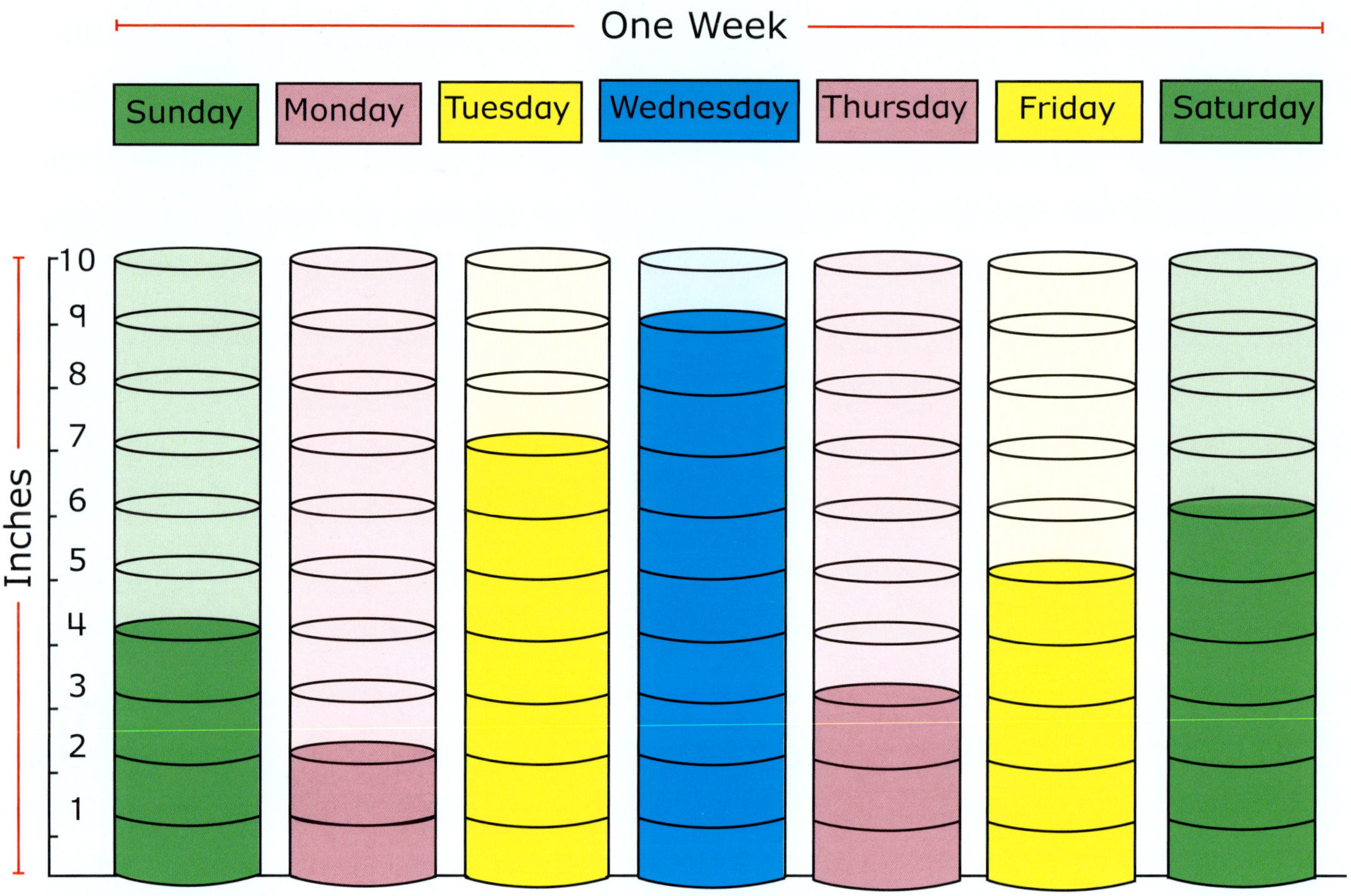

1. These cups show the inches of rain in a rainforest

   in one ____________.

2. What day did it rain the most? ____________

3. What day did it rain the least? ____________

4. What two days did it rain the most? ____________

   ____

5. Did it rain more at the beginning of the week or at the end of the week? ____________ Explain your thinking.

Use the grid to draw a square that is 2 inches high.

Use the grid to draw a square that is 3 inches high.

Clothing Store Sales

1. Circle the item the store sold the most of.
2. Put an X over the item the store sold the least of.
3. Based on what the store has sold, circle all the sentences that are probably true.

   a. The weather in this town is always hot and sunny.

   b. The store sells only men's clothing.

   c. It rains sometimes in this town.

   d. The store sells more rain products than sun products.

Individually subtract the ones and tens columns.

Subtract the following numbers by subtracting the **green ones** column and then the **blue tens** column.

Tens Column → 22 ← Ones Column

```
  22        34        52
- 11      - 21      - 31
----      ----      ----
  11

  88        56        37
- 66      - 33      - 27
----      ----      ----
```

Subtract the following numbers by subtracting the ones column and then the tens column.

```
  34        64        58
- 22      - 22      - 41
----      ----      ----

  68        66        87
- 51      - 33      - 61
----      ----      ----
```

Do the problems. Go in order from left to right. Then, on the next page, connect the answer dots to find the mystery object.

$\begin{array}{r}13\\-12\\\hline\end{array}$	$\begin{array}{r}23\\-21\\\hline\end{array}$	$\begin{array}{r}36\\-24\\\hline\end{array}$	$\begin{array}{r}17\\-14\\\hline\end{array}$
$\begin{array}{r}34\\-12\\\hline\end{array}$	$\begin{array}{r}42\\-21\\\hline\end{array}$	$\begin{array}{r}53\\-22\\\hline\end{array}$	$\begin{array}{r}21\\-10\\\hline\end{array}$
$\begin{array}{r}14\\-11\\\hline\end{array}$	$\begin{array}{r}33\\-12\\\hline\end{array}$	$\begin{array}{r}33\\-21\\\hline\end{array}$	$\begin{array}{r}35\\-14\\\hline\end{array}$
$\begin{array}{r}23\\-22\\\hline\end{array}$	$\begin{array}{r}26\\-24\\\hline\end{array}$	$\begin{array}{r}34\\-24\\\hline\end{array}$	$\begin{array}{r}35\\-24\\\hline\end{array}$
$\begin{array}{r}35\\-13\\\hline\end{array}$	$\begin{array}{r}45\\-35\\\hline\end{array}$	$\begin{array}{r}33\\-21\\\hline\end{array}$	$\begin{array}{r}14\\-14\\\hline\end{array}$

You use me for protection when danger may lie ahead.

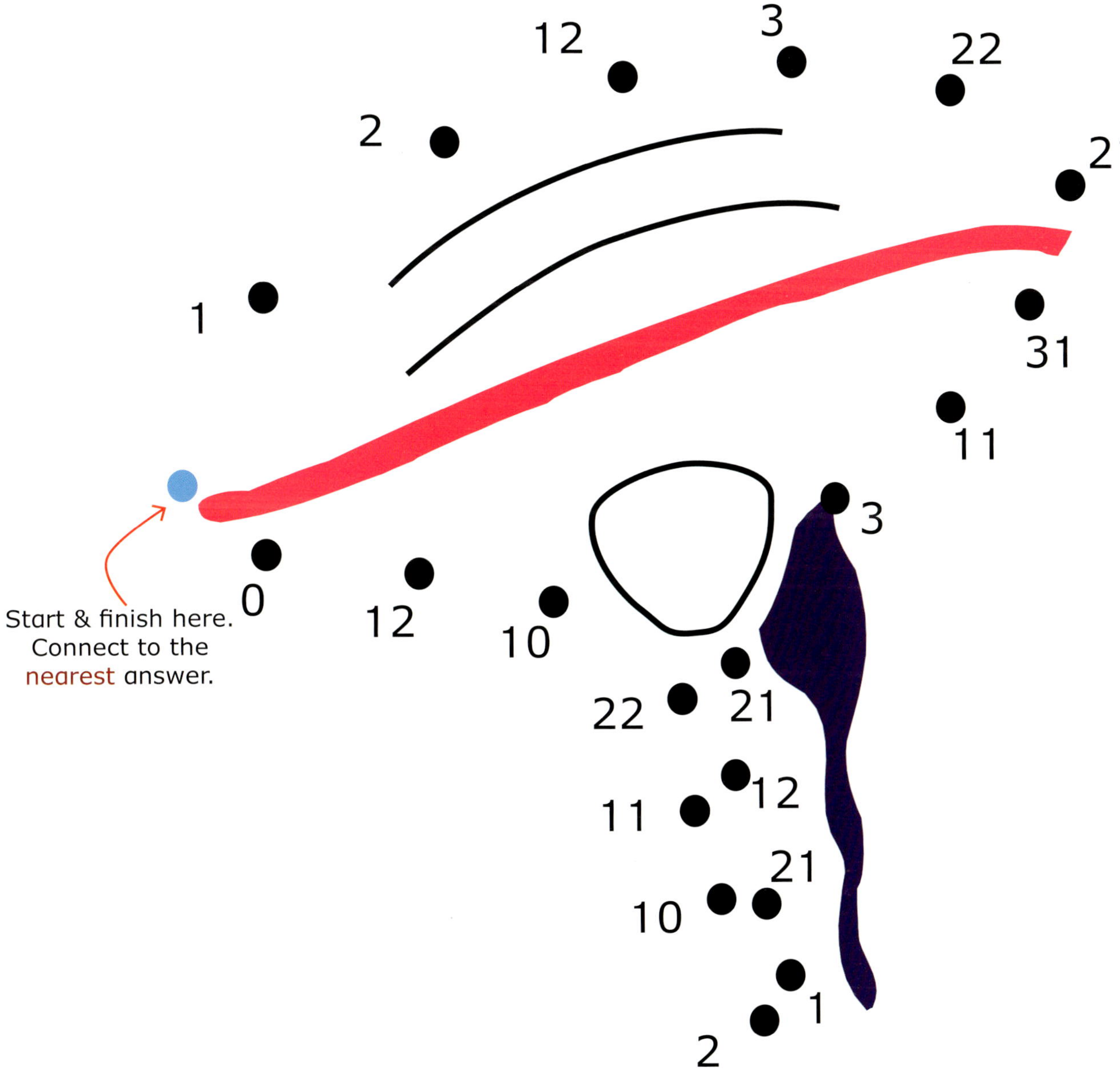

Color the mystery object and add something new to the picture.

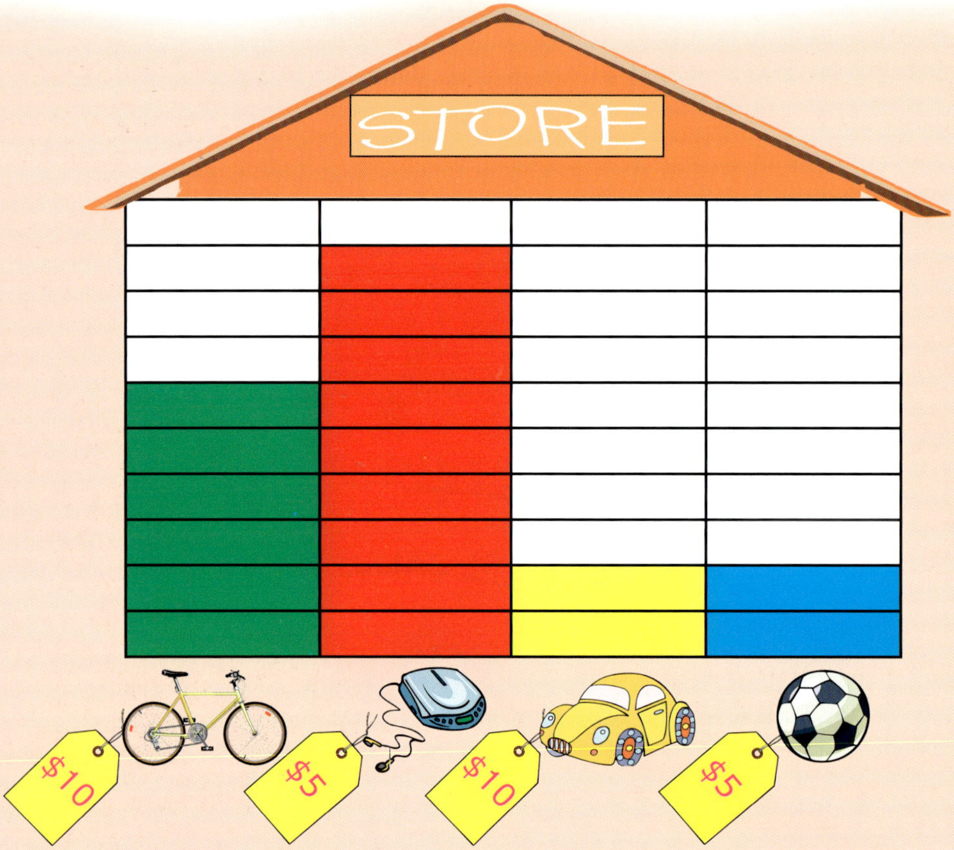

The bar graph shows all the items the store sold today and the price of each item. Answer the questions and explain your thinking.

1. Did the store sell more bicycles or music players?

2. Did the store sell fewer toy cars or soccer balls?

3. How many total dollars was the store paid for its toy cars?

   $20 $10 $50

4. How many total dollars was the store paid for its music players?

   $20 $45 $50

5. Was the store paid more total dollars for bicycles or music players?

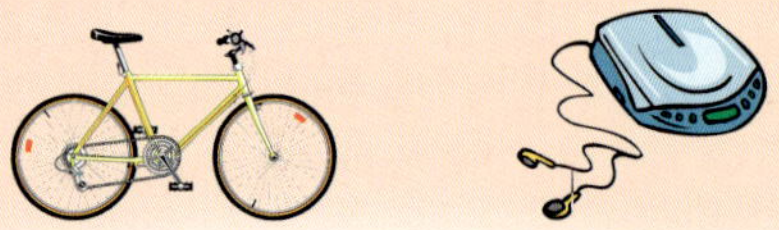

Draw a line of symmetry on each letter. Some letters may have more than one line of symmetry.

U V W

X Y

Which letter has the most lines of symmetry? ________

Which letters have only one line of symmetry? ________

Draw a symmetric figure and mark the line of symmetry.

Draw line segments to match the shape on the left with the shape on the right to make a square.

Circle any flat pattern that could be folded into a cube. Then cut the patterns out and fold them to check your answers after completing the activity on page 240.

One Week

1. Write the name of the day that comes after Tuesday. __________

2. Write the name of the day that comes before Tuesday. __________

3. Write the name of the day that comes 3 days after Sunday. __________

4. Write the name of the day that comes 6 days after Wednesday. __________

5. Write the name of the day that comes a week before Thursday. __________

6. Write the name of the day that comes 1 day before the day after Sunday. __________

7. Write the name of the day that comes 3 days after 3 days before Friday. __________

MARCH

	Sunday	Monday	Tuesday	Wednesday	Thursday	Friday	Saturday
Week 1 →	1	2	3	4	5	6	7
Week 2 →	8	9	10	11	12	13	14
Week 3 →	15	16	17	18	19	20	21
Week 4 →	22	23	24	25	26	27	28
Week 5 →	29	30	31				

APRIL

Sunday	Monday	Tuesday	Wednesday	Thursday	Friday	Saturday
			1	2	3	4
5	6	7	8	9	10	11
12	13	14	15	16	17	18
19	20	21	22	23	24	25
26	27	28	29	30		

1. If today is Saturday, March 7th, what day is next Saturday?

   March ______________.
   (date of the month)

2. If today is Sunday, March 8th, what day will it be in one week?

   ______________, March ______________.
   (day of the week) (date of the month)

3. If today is Tuesday, March 31st, what day is tomorrow?

   ______________, ______________ ______________.
   (day of the week) (month) (date of the month)

4. If today is April 13th, what day will it be in two weeks?

   ______________, ______________ ______________.
   (day of the week) (month) (date of the month)

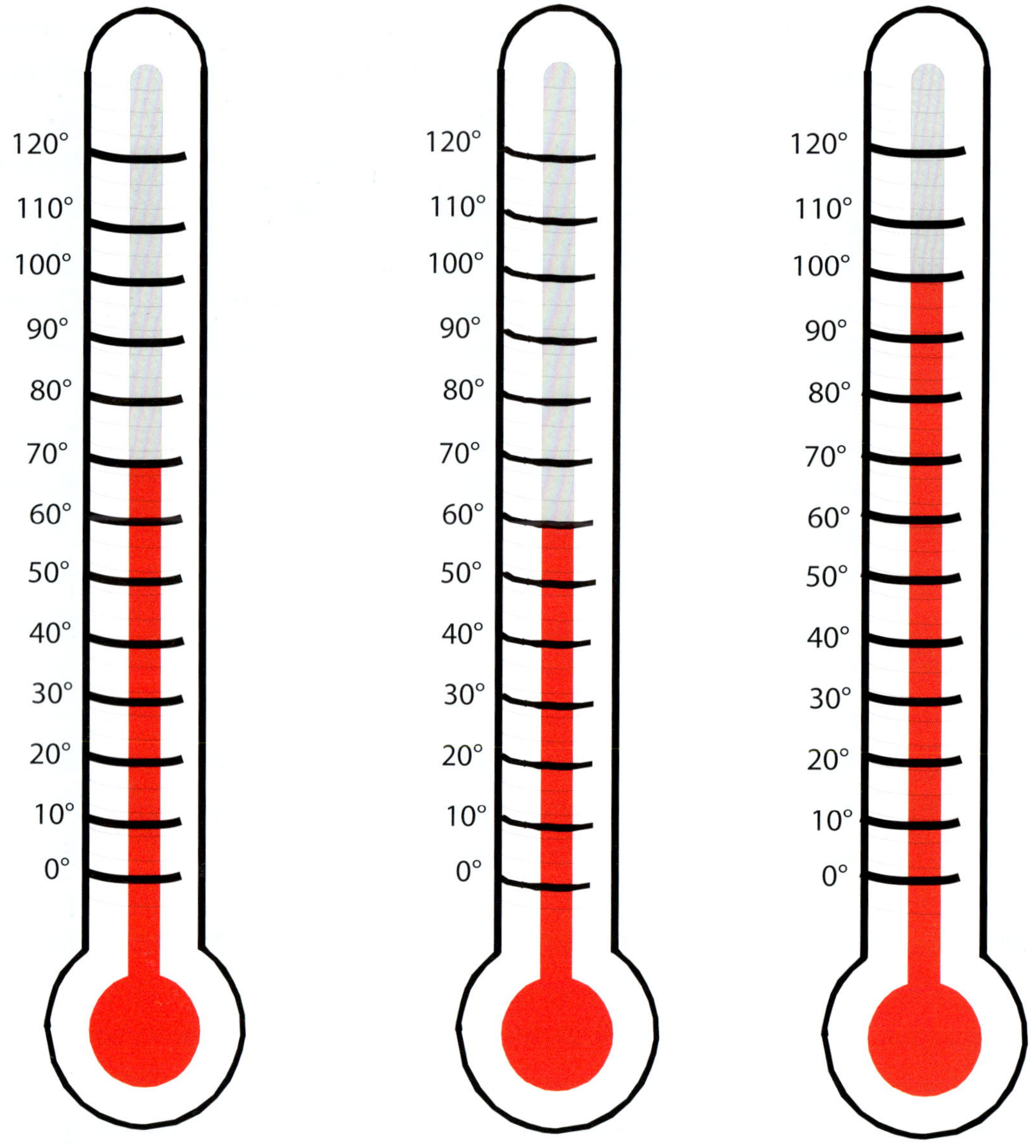

Circle the thermometer that shows the coolest day.

What is the temperature of the coolest day? ________

Explain your answer.

Circle the temperature of the hottest day. ________

Explain your answer.

Monday was hot. The temperature hit one hundred degrees. The next two days, the temperature dropped ten degrees each day. By Wednesday, the temperature was eighty degrees. Write the temperature for each day below and color each thermometer so it shows the correct temperature.

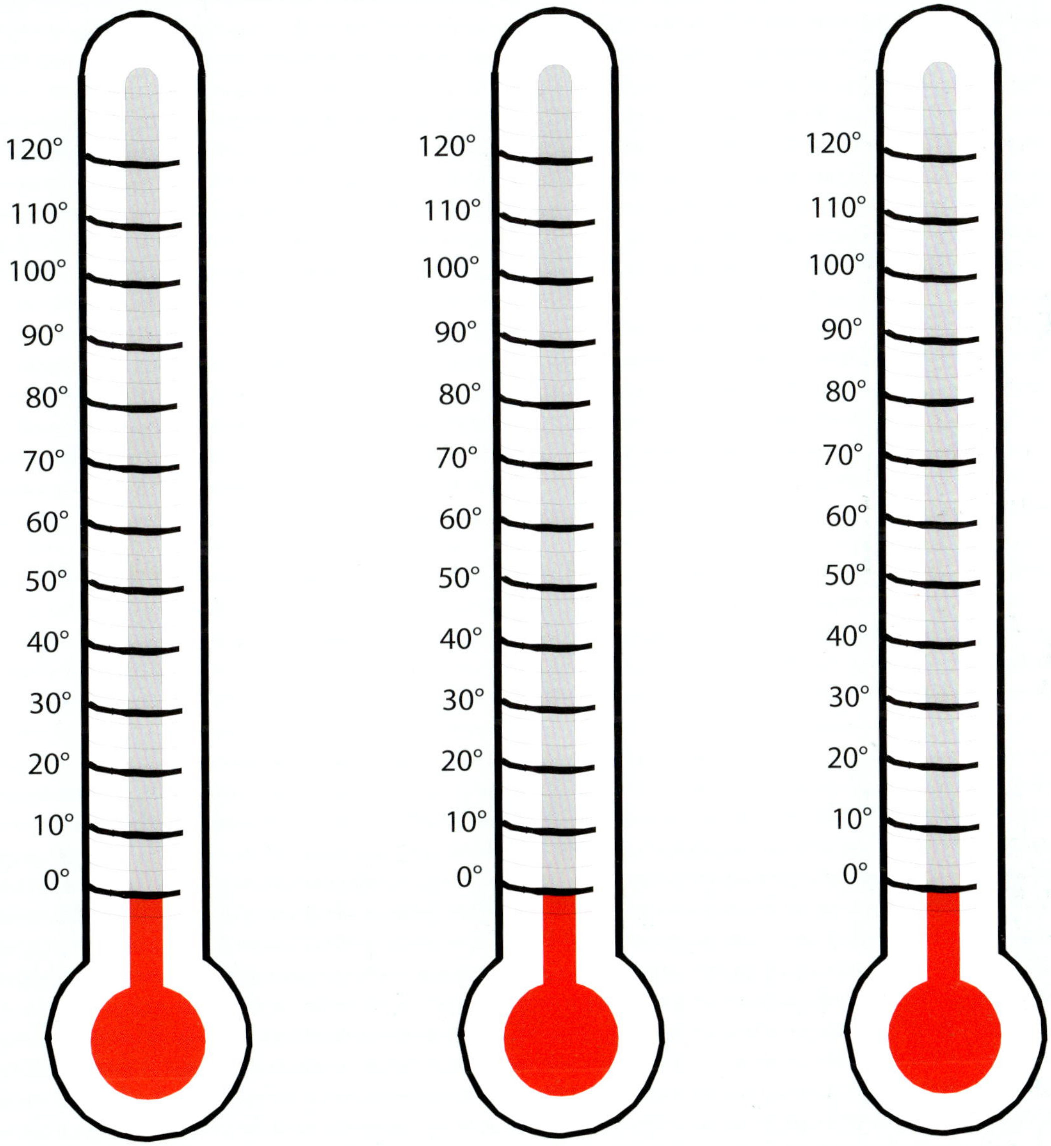

______________	______________	______________
Monday	Tuesday	Wednesday

Do the problems. Go in order from left to right. Then, on the next page, connect the answer dots to find the mystery animal.

68	43	36	77
-24	-21	-14	-54
49	42	73	41
-24	-21	-12	-10
54	83	68	57
-42	-12	-54	-24
65	66	44	45
-34	-11	-22	-44
96	92	42	54
-35	-20	-21	-24

I lived long ago. I was as tall as 70 feet and made the ground shake when I walked.

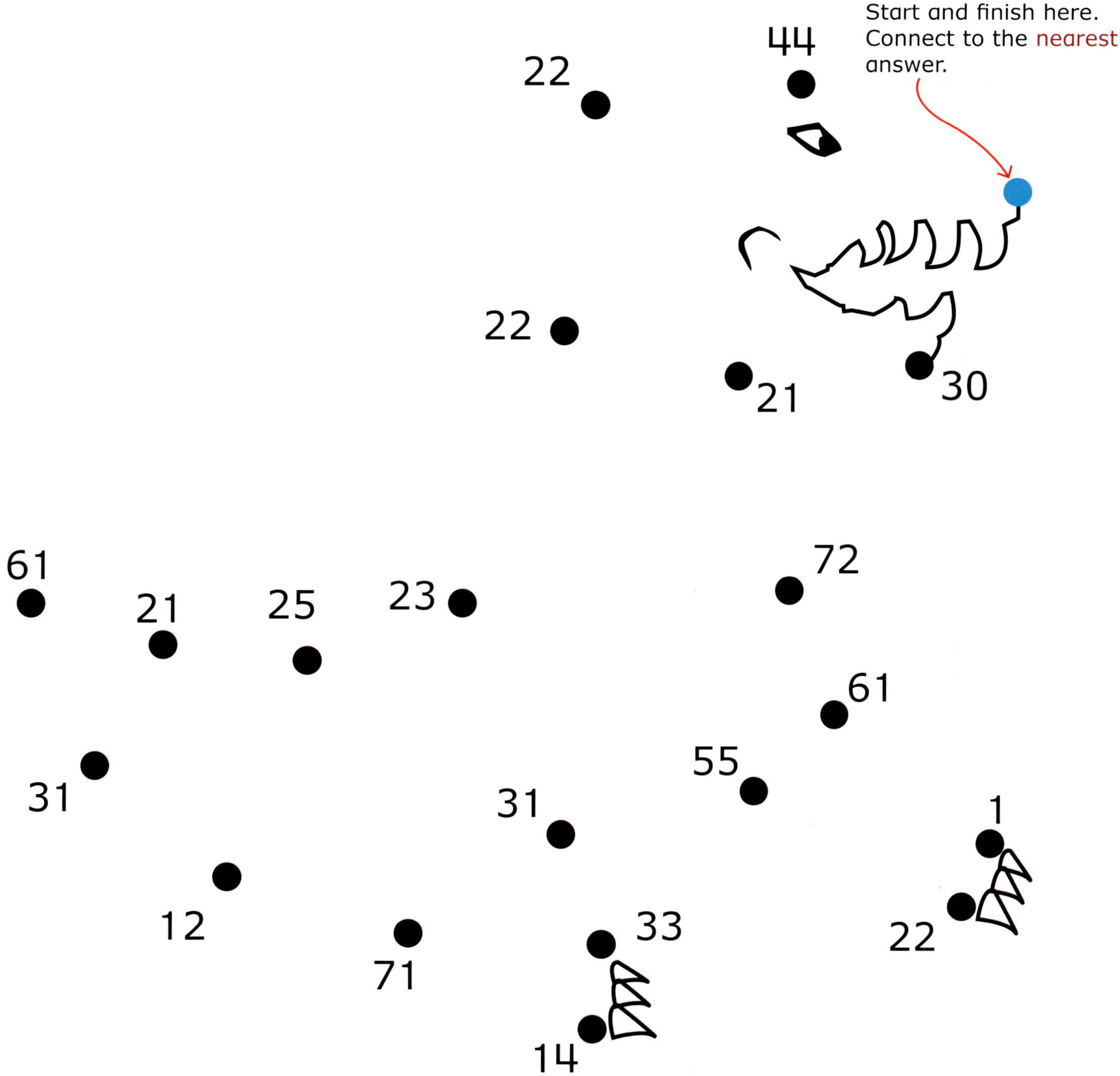

Color the mystery animal and add something new to the picture.

Number Lines do not have to look like this horizontal number line.

For example, locating number 54 on the number line looks like this.

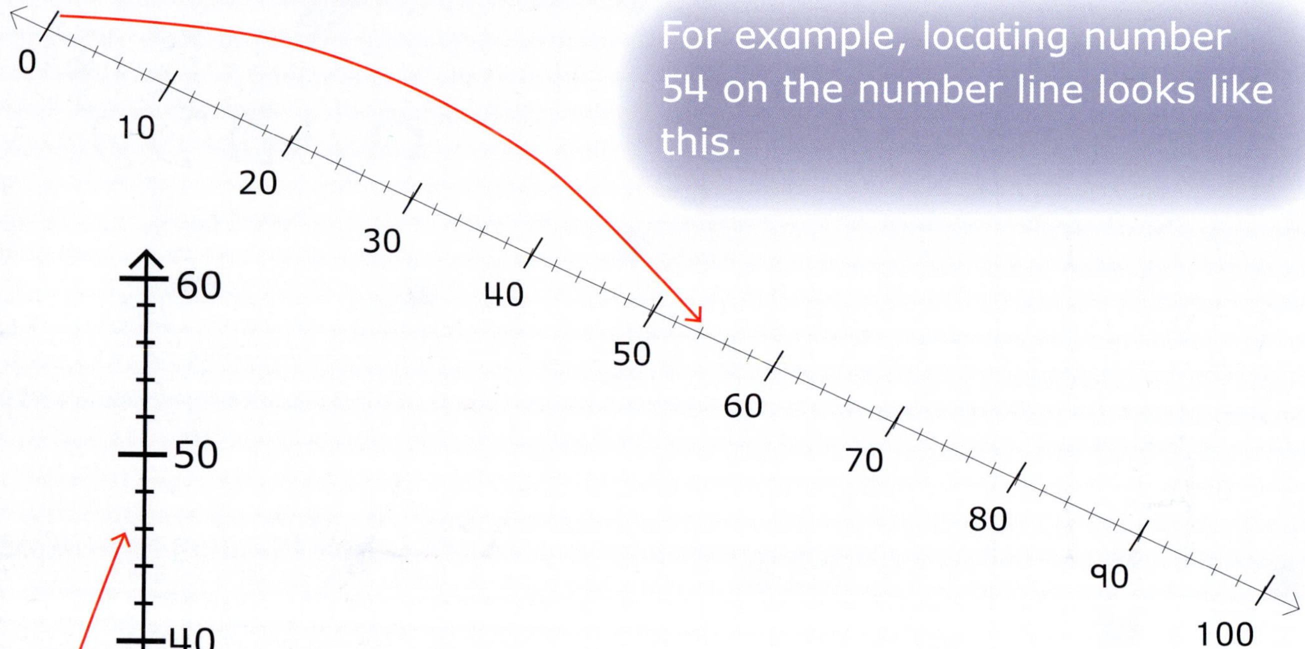

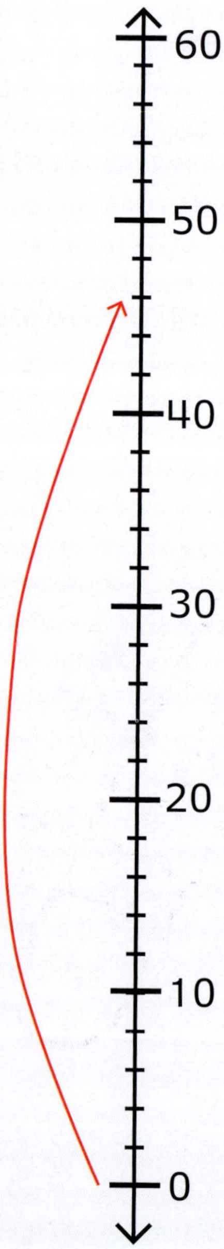

A vertical line looks like this. What number is shown on this number line?

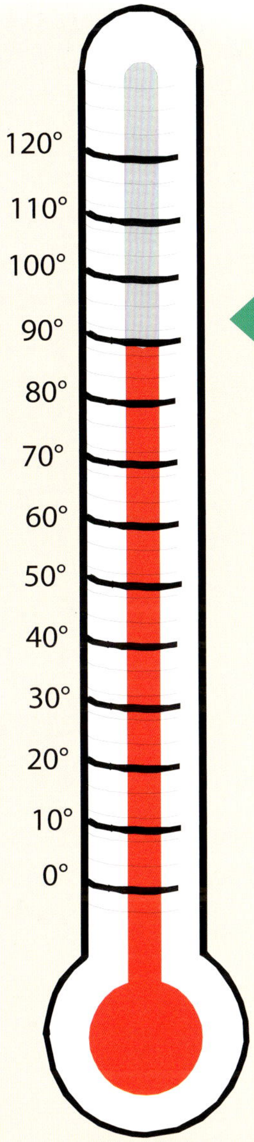

What is the temperature of the day? ______

If you are outside, would you want a warm or cold drink? _____

Explain your answer.

What is the temperature of the day? ______

If you are outside, would you want a warm or cold drink?

_____

Explain your answer.

Trace the numerals and do the problems. Go in order from left to right. Then, on the next page, connect the answer dots to find the mystery animal.

12 - 2 = ______	35 - 2 = ______
48 - 7 = ______	69 - 8 = ______
88 - 7 = ______	56 - 6 = ______
97 - 3 = ______	78 - 6 = ______
33 - 2 = ______	56 - 5 = ______
22 - 1 = ______	19 - 5 = ______
36 - 4 = ______	55 - 3 = ______
58 - 5 = ______	47 - 6 = ______
79 - 7 = ______	28 - 8 = ______
39 - 9 = ______	44 - 0 = ______

I always wear a coat and clean it every day.

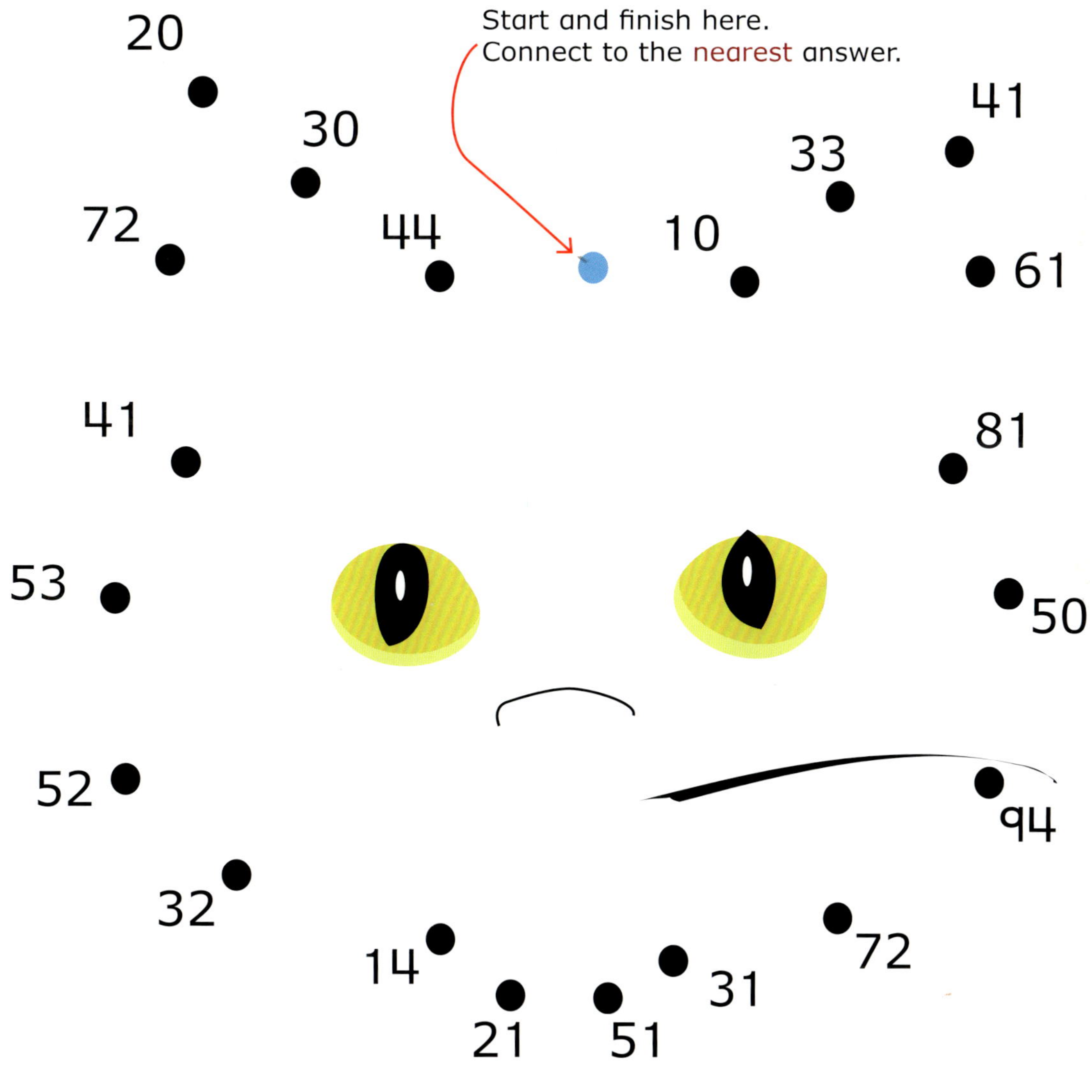

Color the mystery animal and add something new to the picture.

Circle the shapes in each set that are needed to make each shape in blue.

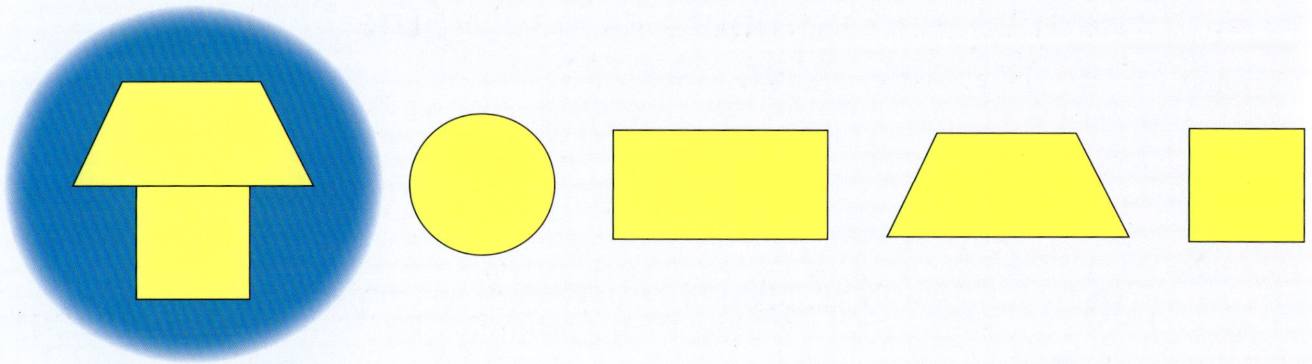

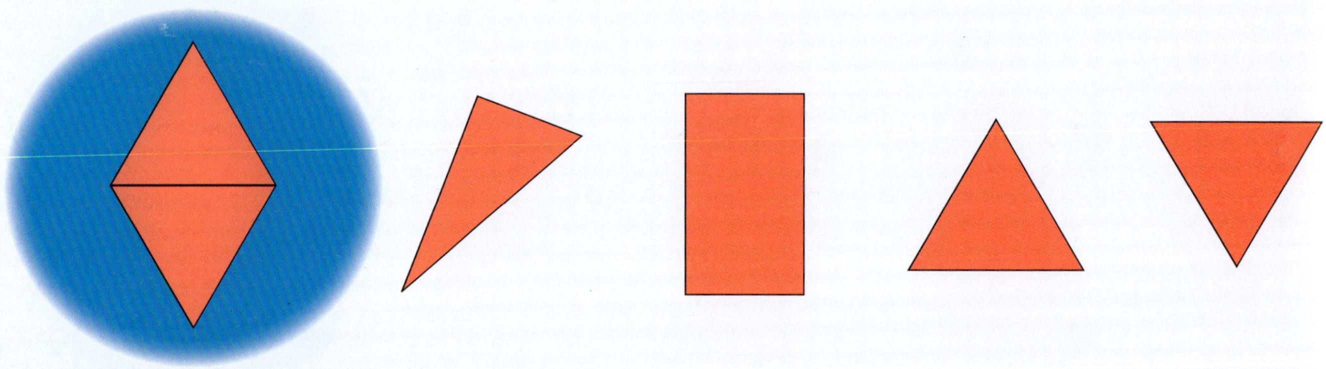

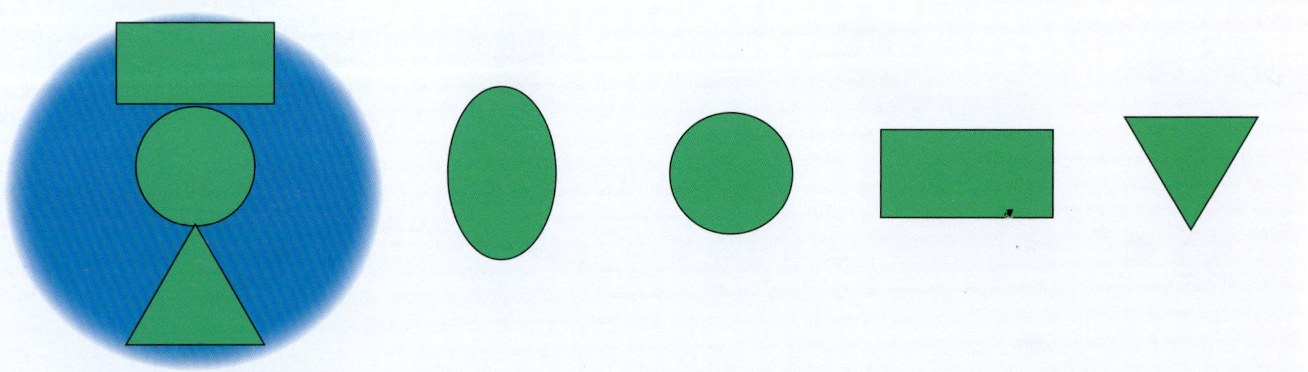

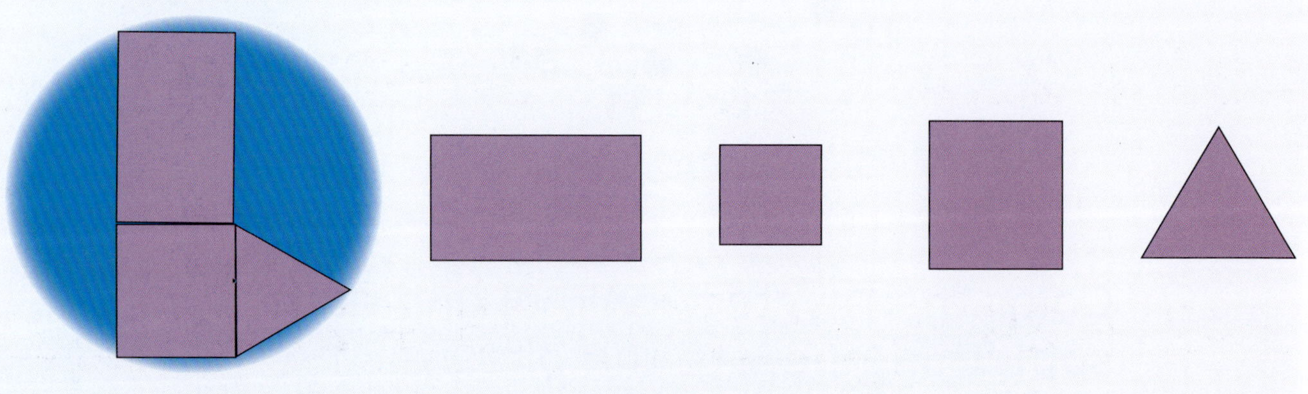

NOTE: For students who struggle with this page, please see page v of the Introduction.

Individually add the ones and tens columns.

Add the following numbers by adding the **green ones** and then the **blue tens**.

Tens Ones

11 + 11 = 22

23 + 11 = ____

31 + 14 = ____

9 + 22 = ____

Add the following numbers by adding the ones and then the tens.

11 + 21 = ____

12 + 21 = ____

32 + 23 = ____

62 + 36 = ____

42 + 24 = ____

73 + 25 = ____

Trace the numerals and do the problems. Go in order from left to right. Then, on the next page, connect the answer dots to find the mystery animal.

31 + 52 = ______ 43 + 24 = ______

53 + 41 = ______ 67 + 12 = ______

36 + 43 = ______ 71 + 21 = ______

37 + 12 = ______ 91 + 1 = ______

54 + 45 = ______ 42 + 24 = ______

36 + 63 = ______ 63 + 16 = ______

72 + 27 = ______ 14 + 25 = ______

23 + 25 = ______ 23 + 15 = ______

32 + 34 = ______ 54 + 22 = ______

27 + 51 = ______ 43 + 35 = ______

I get up very early in the morning and make a loud noise to wake up everyone.

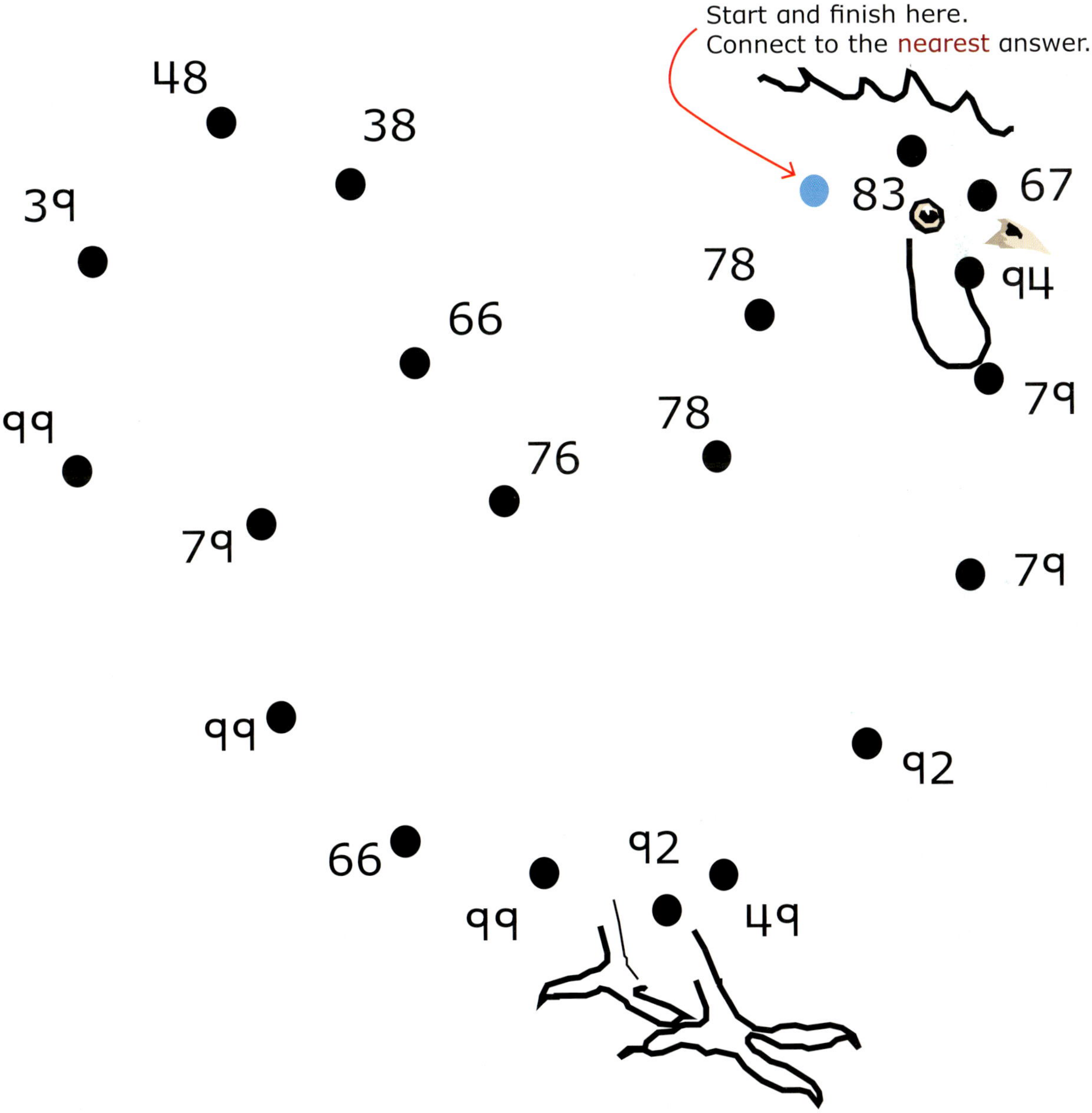

Color the mystery animal and add something new to the picture.

Circle the shapes in each set that are needed to make each shape in pink.

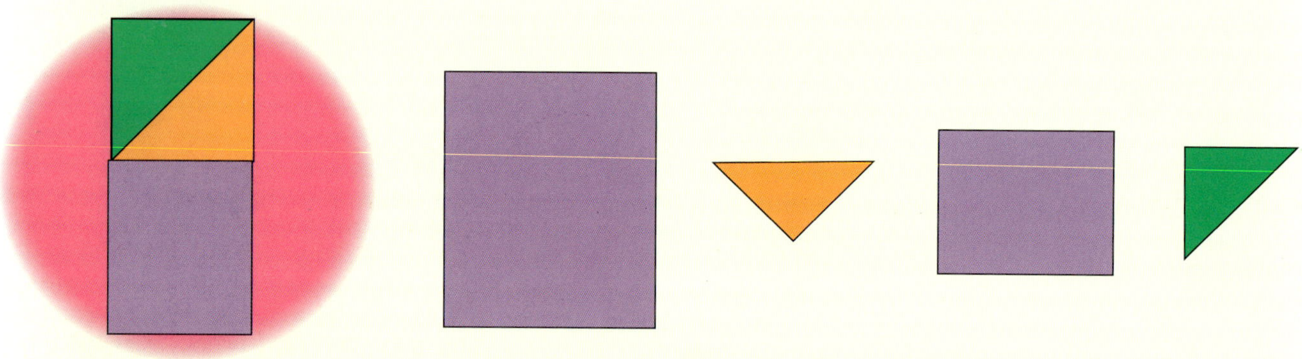

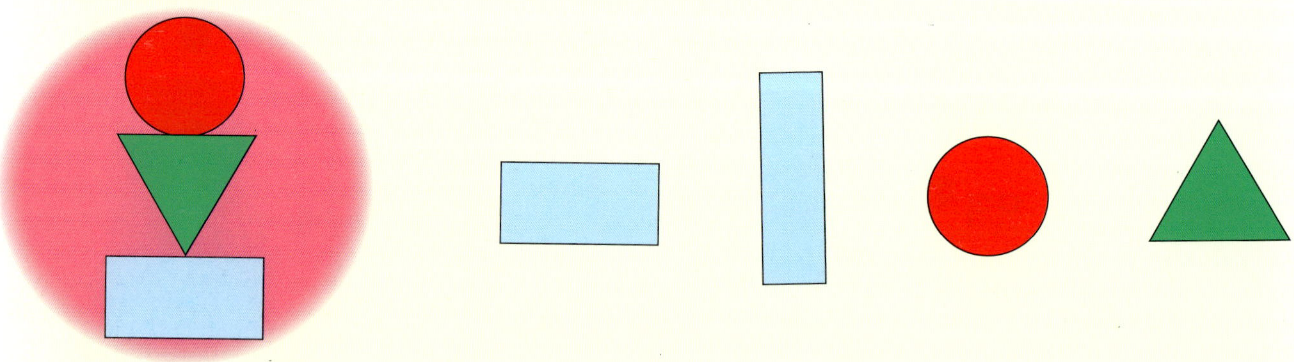

* NOTE: For students who struggle with this page, please see page v of the Introduction.

10¢ + 3¢ = 13¢

Start with a dime and find three pennies. How much money will there be?

5¢ + 20¢ = ____________

Write the amount of money that one nickel and two dimes equal.

15¢ - 9¢ = ____________

Start with a dime and a nickel and then spend nine cents. How much money will there be?

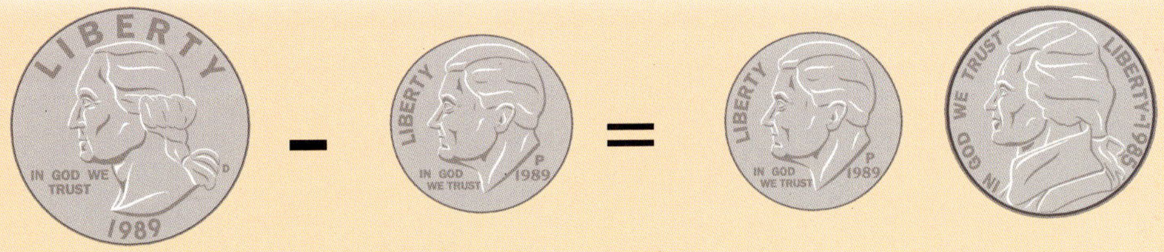

25¢ - 10¢ = ____________

Start with a quarter and spend ten cents. Write the amount of money left.

Trace the numerals and do the problems. Go in order from left to right. Then, on the next page, connect the answer dots to find the mystery animal.

INSTRUCTIONS

76 - 23 = ______ 58 - 27 = ______

86 - 74 = ______ 99 - 88 = ______

85 - 43 = ______ 67 - 25 = ______

96 - 36 = ______ 78 - 61 = ______

39 - 24 = ______ 56 - 11 = ______

47 - 20 = ______ 97 - 83 = ______

69 - 51 = ______ 56 - 35 = ______

84 - 51 = ______ 72 - 31 = ______

86 - 72 = ______ 95 - 53 = ______

96 - 41 = ______ 48 - 10 = ______

My noise is wet and cold. I can smell food better than you.

Start and finish here.
Connect to the nearest answer.

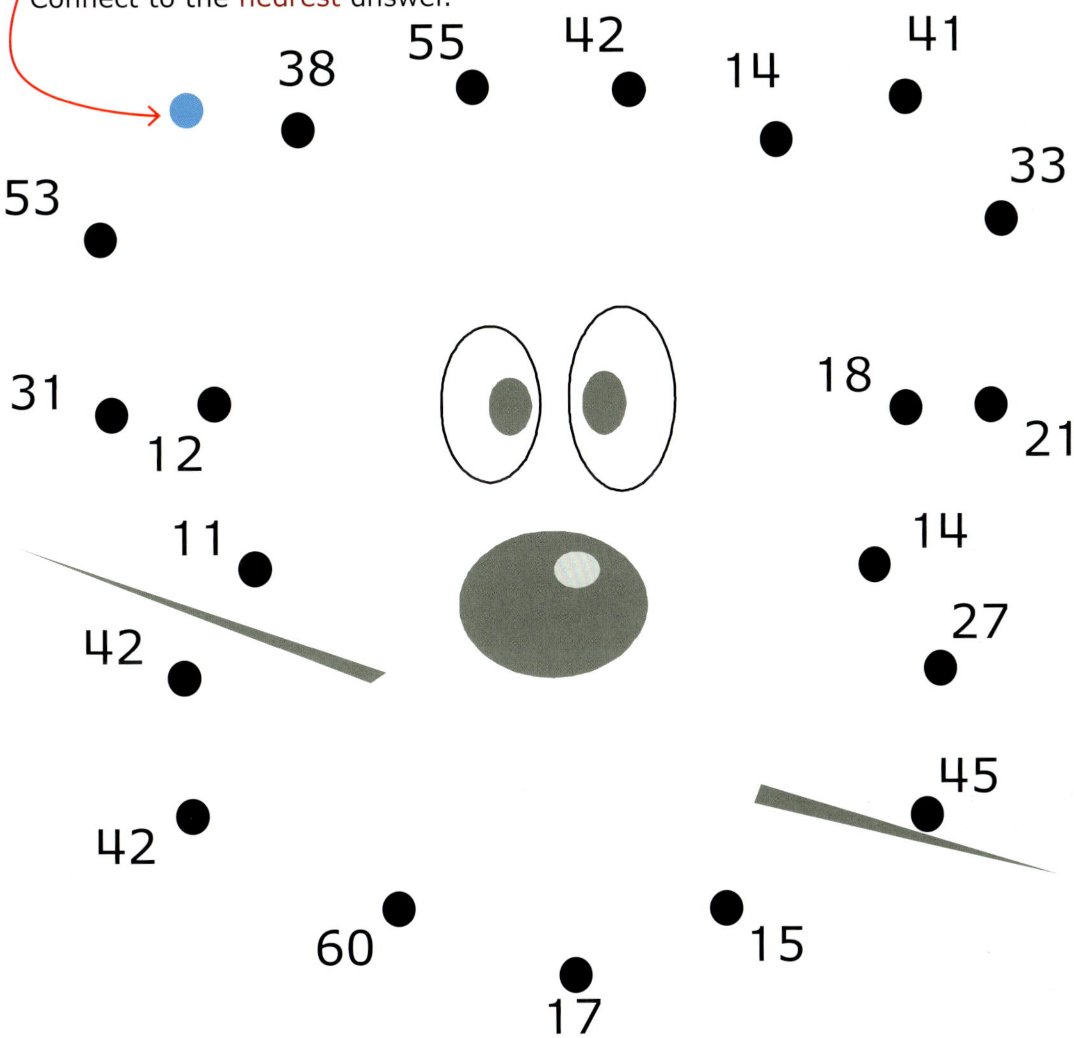

Color the mystery animal and add something new to the picture.

Write and say the temperature of each thermometer. Then put an X on the thermometer that shows the warmer temperature.

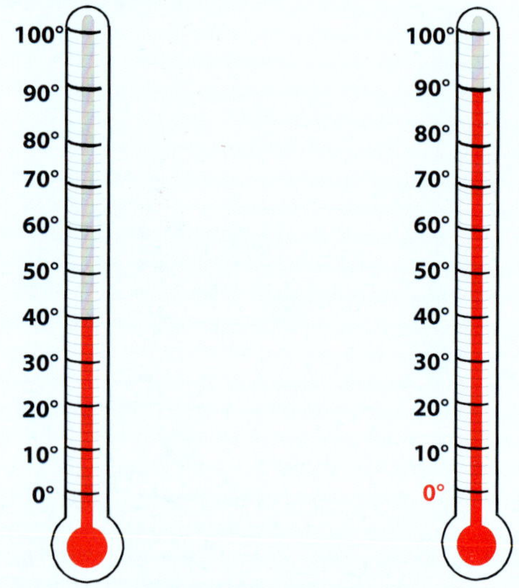

Circle the temperature you estimate for this picture.

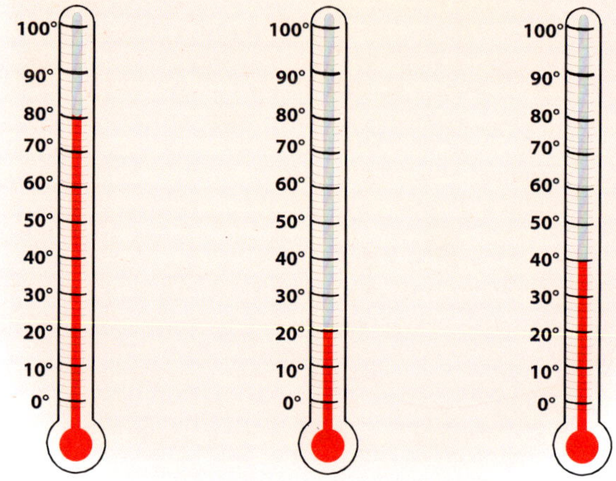

Circle the temperature you estimate for this picture.

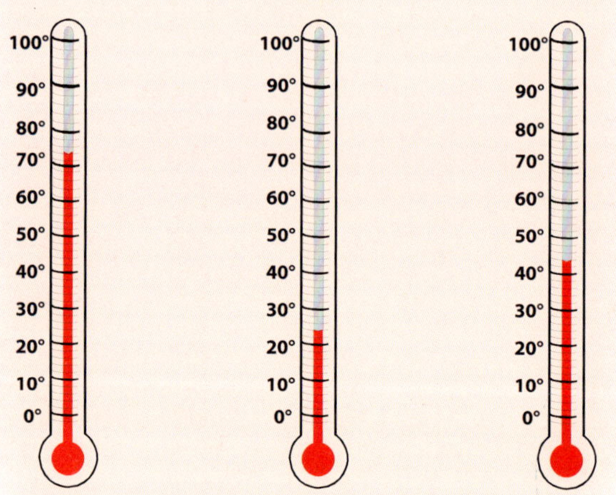

Circle the temperature you estimate for this picture.

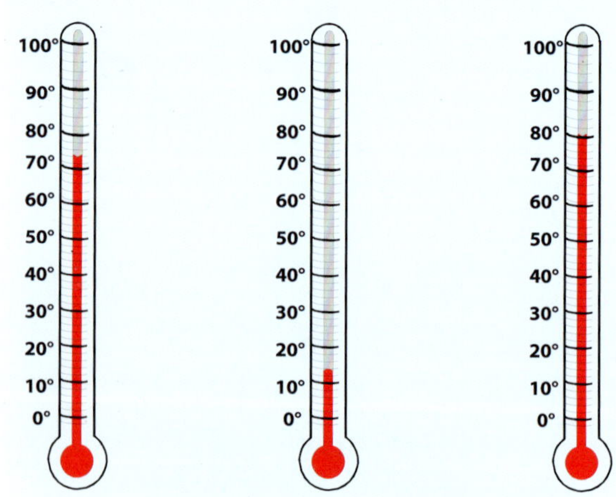

Circle any flat pattern that could be folded into a cube. Then cut the patterns out and fold them to check your answers after completing the activity on page 260.

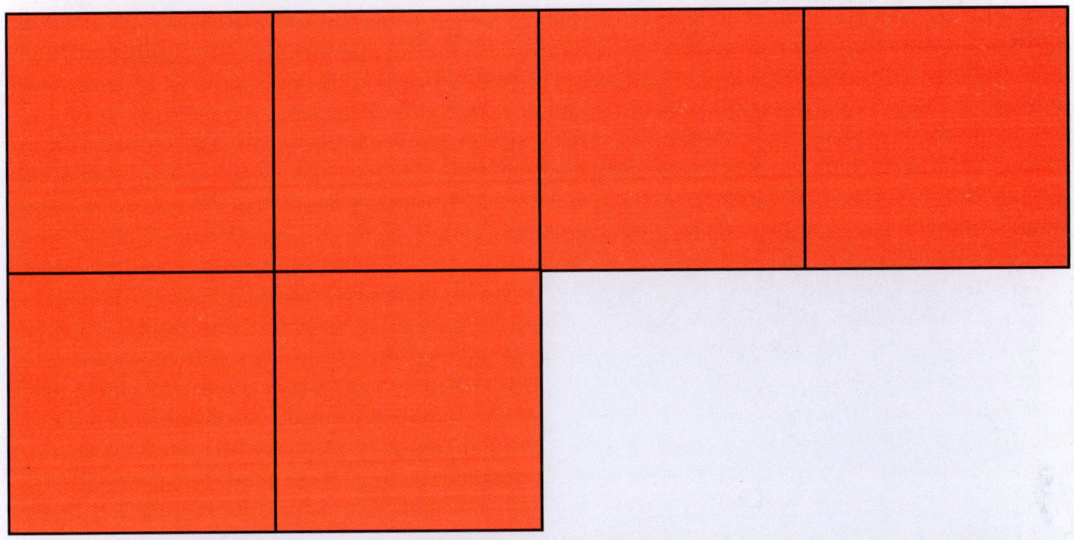

Describe the danger in each picture.

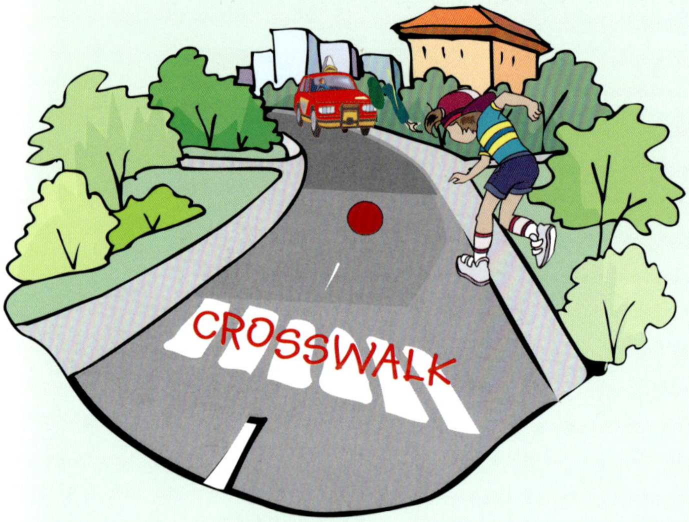

Circle any flat pattern that could be folded into a cube. Then cut the patterns out and fold them to check your answers after completing the activity on page 262.

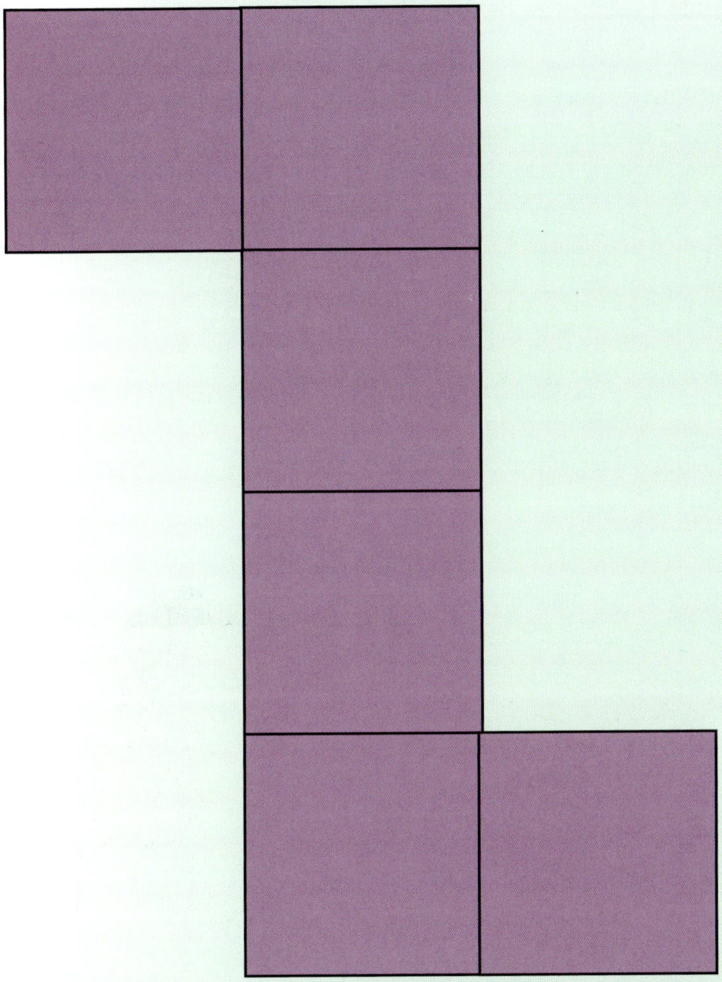

Use the base 10 blocks to solve each problem. Then write and say each number sentence.

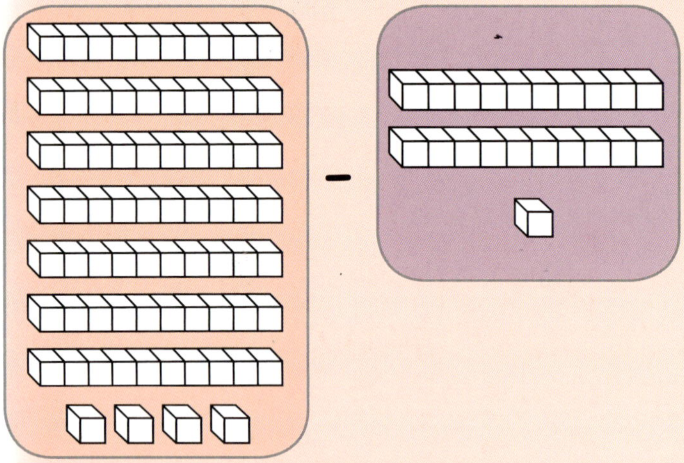

_____ _____ = _____

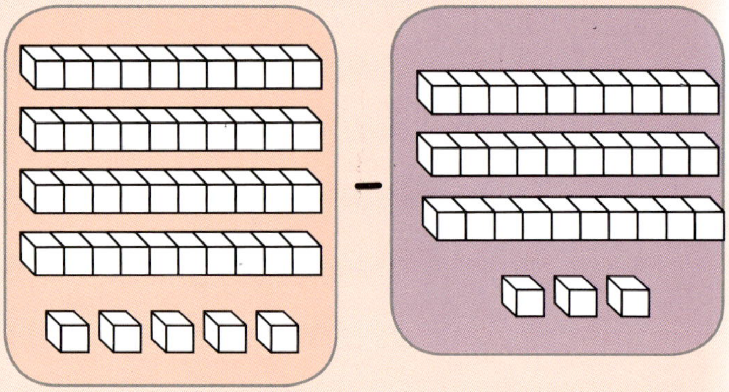

_____ _____ = _____

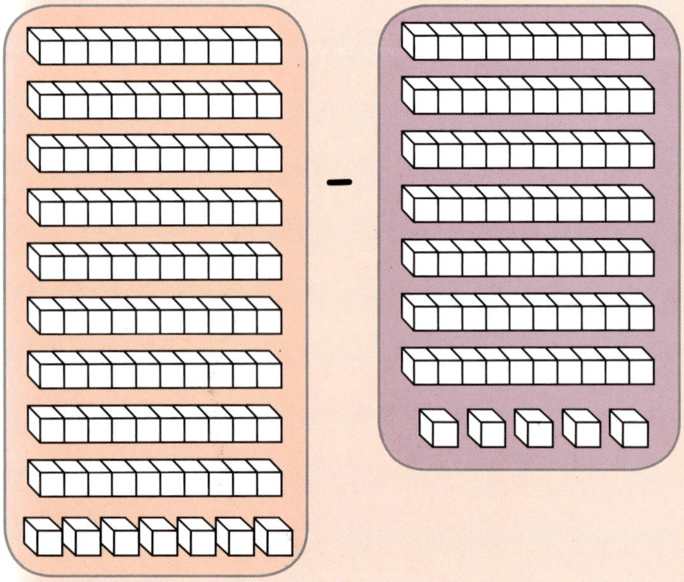

_____ _____ = _____

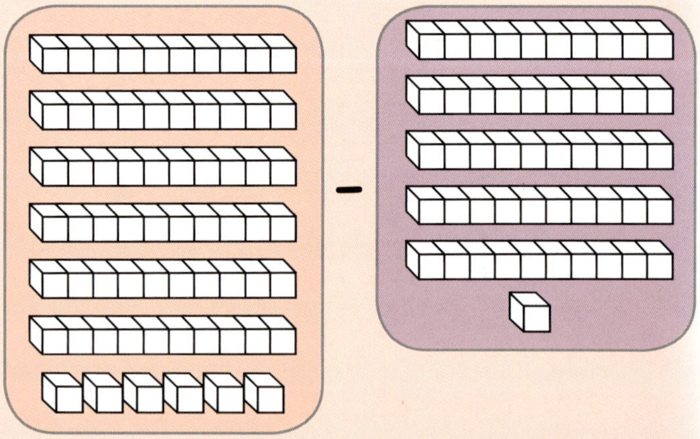

_____ _____ = _____

What time is it on the clock?

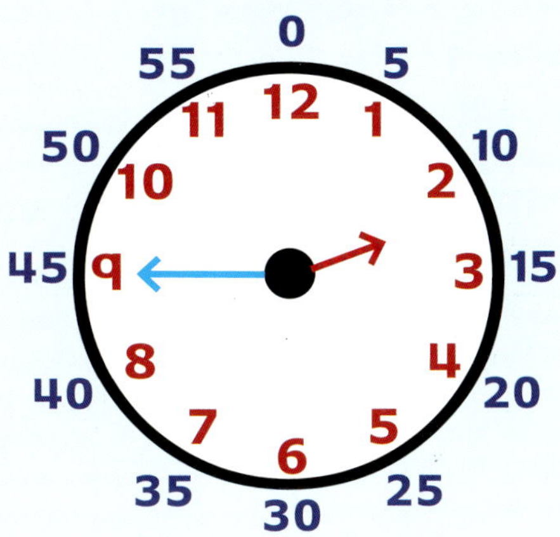

_______ : _______
hour minutes

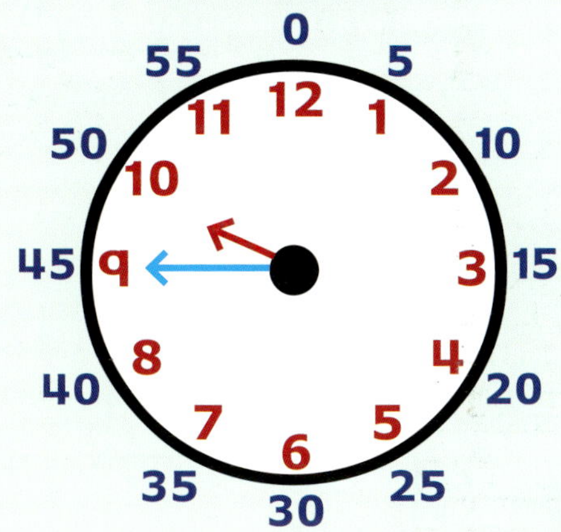

_______ : _______
hour minutes

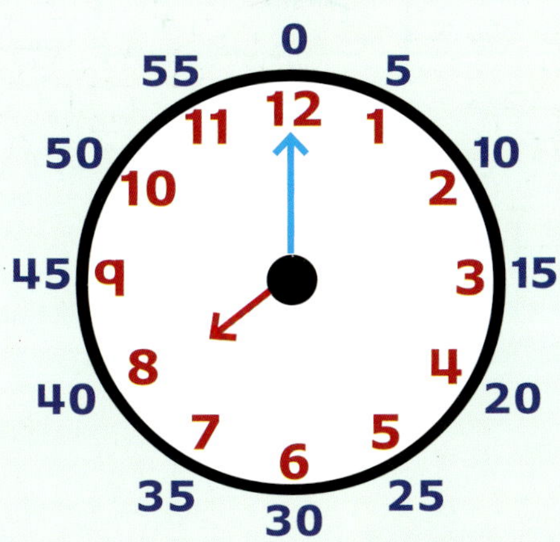

_______ : _______
hour minutes

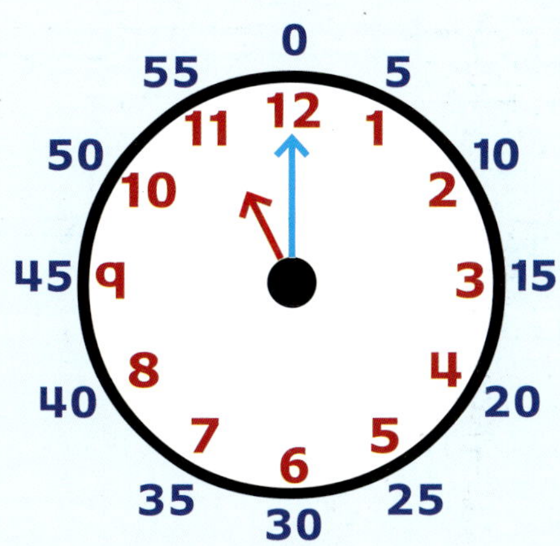

_______ : _______
hour minutes

MIND BENDERS® *

DIRECTIONS: Fill in the chart using Y for yes or N for no as you solve the puzzle.

A boy, girl, mom, and dad all fell asleep at different times. Find out when each person fell asleep.

1. The boy was asleep after his sister and dad.

2. The mom was asleep after the boy.

3. The dad was asleep before 9:00.

* For more Mind Benders® activities, please see our *Mind Benders®* series.